Nuclear Reactor Safety

NUCLEAR SCIENCE AND TECHNOLOGY

A Series of Monographs and Textbooks

CONSULTING EDITOR

V. L. PARSEGIAN

Chair of Rensselaer Professor
Rensselaer Polytechnic Institute
Troy, New York

1. John F. Flagg (Ed.)
   CHEMICAL PROCESSING OF REACTOR FUELS, 1961
2. M. L. Yeater (Ed.)
   NEUTRON PHYSICS, 1962
3. Melville Clark, Jr., and Kent F. Hansen
   NUMERICAL METHODS OF REACTOR ANALYSIS, 1964
4. James W. Haffner
   RADIATION AND SHIELDING IN SPACE, 1967
5. Weston M. Stacey, Jr.
   SPACE-TIME NUCLEAR REACTOR KINETICS, 1969
6. Ronald R. Mohler and C. N. Shen
   OPTIMAL CONTROL OF NUCLEAR REACTORS, 1970
7. Ziya Akcasu, Gerald S. Lellouche, and Louis M. Shotkin
   MATHEMATICAL METHODS IN NUCLEAR REACTOR DYNAMICS, 1971
8. John Graham
   FAST REACTOR SAFETY, 1971
9. Akinao Shimizu and Katsutada Aoki
   APPLICATION OF INVARIANT EMBEDDING TO REACTOR PHYSICS, 1972
10. Weston M. Stacey, Jr.
    VARIATIONAL METHODS IN NUCLEAR REACTOR PHYSICS, 1974
11. T. W. Kerlin
    FREQUENCY RESPONSE TESTING IN NUCLEAR REACTORS, 1974
12. F. R. Farmer
    NUCLEAR REACTOR SAFETY, 1977

Nuclear Reactor Safety

Edited by
F. R. FARMER
United Kingdom Atomic Energy Authority
Warrington, Cheshire, England

ACADEMIC PRESS   New York   San Francisco   London   1977

A Subsidiary of Harcourt Brace Jovanovich, Publishers

COPYRIGHT © 1977, BY ACADEMIC PRESS, INC.
ALL RIGHTS RESERVED.
NO PART OF THIS PUBLICATION MAY BE REPRODUCED OR
TRANSMITTED IN ANY FORM OR BY ANY MEANS, ELECTRONIC
OR MECHANICAL, INCLUDING PHOTOCOPY, RECORDING, OR ANY
INFORMATION STORAGE AND RETRIEVAL SYSTEM, WITHOUT
PERMISSION IN WRITING FROM THE PUBLISHER.

ACADEMIC PRESS, INC.
111 Fifth Avenue, New York, New York 10003

United Kingdom Edition published by
ACADEMIC PRESS, INC. (LONDON) LTD.
24/28 Oval Road, London NW1

Library of Congress Cataloging in Publication Data

Main entry under title:

Nuclear reactor safety.

(Nuclear science and technology series; v. 12)
Includes bibliographical references.
1. Nuclear reactors–Safety measures. I. Farmer,
Frank Reginald, Date
TK9152.N795   621.48'35   76-58008
ISBN 0–12–249350–8

PRINTED IN THE UNITED STATES OF AMERICA
79 80 81 82   9 8 7 6 5 4 3 2

Contents

List of Contributors   ix

Preface   xi

1 Introduction   1
    H. M. Nicholson

2 Radioactivity and the Fission Products   5
    F. Abbey

    Introduction   5
    Radioactivity and Its Effects   6
    Fission Product Quantities and Relative Importance   8
    The Chemical Form of Fission Products and Their Behavior in Normal
       Reactor Operation   13
    Fission Product Behavior in Accident Conditions   18
    References   28

3 Radiation Hazards and Environmental Consequences of Reactor Accidents   31
    J. R. Beattie

    Radiation Hazards and Health Physics Control Levels   31
    Reactor Accidents and Fission Product Release to the Atmosphere   37
    References   47

4 The Calculated Risk—A Safety Criterion
G. D. Bell
49

References … 71

5 Quantitative Approach to Reliability of Control and Instrumentation Systems
A. Aitken
73

Introduction … 73
Reliability and Probability … 73
Equipment Failure Rates … 74
Typical Systems … 77
Analysis of Protective Systems … 78
Reliability Assessment of Protective Equipment … 86
General Methods of Assessment of Overall Performance … 105
References … 107

6 The Reliability of Heat Removal Systems
F. M. Davies
109

Introduction … 109
Heat Sources … 110
Reliability Targets … 111
Reliability Evaluation … 112
Reactor Systems … 118
Decay Heat Rejection Capacity … 125
Reliability Analysis of Thermal Syphon Loops … 125
Logic Diagram … 126
Criterion for System Failure … 130
Analysis Using NOTED Program … 131
Results … 133
Common Fault Modes … 135
Pressurized Reactors … 138
Discussion … 140
Appendix. Mathematical Models … 141
References … 145

7 The Integrity of Pressure Vessels
R. O'Neil
149

Introduction … 149
Requirements … 150
Statistical Evidence … 152
Relevant Experience of Nuclear Pressure Vessel Integrity … 153

Contents

Experience and Estimates Based on Nonnuclear Pressure Vessels	153
Summary of Relevant Nonnuclear Experience	158
Extrapolation to Nuclear Vessels	158
Discussion	163
Conclusions	164
References	165

8 Thermal Reactor Safety   167
J. H. Bowen

Fission Product Releases	171
Reactivity Faults	172
Loss-of-Flow Faults	176
Loss-of-Pressure Faults	178
References	182

9 Safety of Fast Reactors   183
H. J. Teague

Special Features Important to Safety in Fast Reactors	188
Sources of Hazard	192
Faults Due to Local Loss of Cooling in a Subassembly	195
Fuel Handling Faults	199
Containment	201
The Effect of a "Hypothetical" Severe Accident on Containment	201
Summary	206
References	207

Index   211

List of Contributors

Numbers in parentheses indicate the pages on which the authors' contributions begin.

F. ABBEY (5), Safety and Reliability Directorate, United Kingdom Atomic Energy Authority, Warrington, Cheshire, England

A. AITKEN (73), Safety and Reliability Directorate, United Kingdom Atomic Energy Authority, Warrington, Cheshire, England

J. R. BEATTIE (31), Safety and Reliability Directorate, United Kingdom Atomic Energy Authority, Warrington, Cheshire, England

G. D. BELL (49), Safety and Reliability Directorate, United Kingdom Atomic Energy Authority, Warrington, Cheshire, England

J. H. BOWEN (167), Safety and Reliability Directorate, United Kingdom Atomic Energy Authority, Warrington, Cheshire, England

F. M. DAVIES (109), Safety and Reliability Directorate, United Kingdom Atomic Energy Authority, Warrington, Cheshire, England

H. M. NICHOLSON (1), Green Close Brook, Newport, Isle of Wight, England

R. O'NEIL (149), Safety and Reliability Directorate, United Kingdom Atomic Energy Authority, Warrington, Cheshire, England

H. J. TEAGUE (183), Safety and Reliability Directorate, United Kingdom Atomic Energy Authority, Warrington, Cheshire, England

Preface

Safety considerations dominated the early development of nuclear power, and the record of the first Geneva Conference on the Peaceful Uses of Atomic Energy provides a background against which the progress in safety criteria can be followed.

Three themes had been well established by 1955:

(a) "One of the current difficulties in evaluating reactor hazards is this lack of experience with reactor accidents [1]." "All other engineering technologies have advanced not on the basis of their successes, but on the basis of their failures. . . . Atomic energy, however, must forego this advantage of progressing on the basis of knowledge gained by failures [2]."

(b) "It is evident that radioactive poisons are more hazardous than chemical poisons by a factor of something like 10^6 to 10^9 [1]."

(c) The potential for harm to people and property is very great for nuclear installations. Parker [3] estimates a full release from a 1000 MWt reactor could kill 200 to 500 people in a region with a population density of 200 to 500 people per square mile. The WASH-740 report estimates that there could be a remote possibility of killing 3400 people [4].

As these ideas developed, various proposals were put forward as aids to safety, such as: safety by siting, by containment, and by the consideration of and protection from the maximum credible accident. Sites for the first phase of the British nuclear power program were chosen within population limits stemming from the Marley and Fry accident analysis; namely, that in any

10° sector there should be less than 500 people within 1½ miles, less than 10,000 people within 5 miles, or less than 100,000 people within 10 miles [5].

The difficulties in the use of precise limits became more and more apparent and this led to changes which were first introduced in an IAEA panel discussion in 1961 [6] and at a CNEN conference in 1962 [7]. In the following years the UK adopted a graduated measure of assessing population devised to find sites having relatively low population near the reactor.

Meanwhile, in the USA from 1959 onward, siting criteria were based on dose limits to the population which were assessed as a result of a maximum credible accident. The calculation model was prescribed and included the benefit of containment, having a leak rate of one part per thousand per day and assumed the release of 50% of the volatile fission products from the fuel. These assumptions seemed largely independent of the real state of the reactor or its containment and supported the belief that safety was achieved by siting.

During this period in the development of safety ideas, the concept of maximum credible accident played a major part, even if in a somewhat confused fashion:

> Thus, the maximum credible accident is defined as the upper limit of hazard, i.e. fission product release, against which the features of the site must be compared [8].

During the period of the mid 1960s, there was a move to bring reactors closer to populations; this required an amendment of the maximum credible accident to become more realistic and required more confidence in the functioning of safety equipment:

> . . . consequences-limiting safeguards *must* have a very high degree of dependability . . . . In summary, "high dependability" in safeguards means systems designed with unreserved commitment to the concept that their effective performance at any time, under any conditions, when they may be called on, is a matter of life and death [8].

There is, at this stage, a move away from the simple concept of "safe" and "unsafe"; a slow move to the recognition of risk or conversely the recognition of the need for a high degree of dependability or reliability. There is still, however, a great reluctance to be in any way specific with regard to the reliability criteria or the acceptance of risk.

By 1964 we were moving slowly and hopefully into the development of realistic safety criteria:

> It was recognized that the limitations placed on a reactor at the design stage could not forever remain arbitrary but should be related to real and recognizable criteria which could be tested [9].

Part of the realism led to the acceptance of some risk:

> The objective of the temperature control criterion is to avoid widespread can melting during the temperature transient following bottom duct failure. It is assumed acceptable to have some low probability of ignition of a single channel, or small number of channels [10]. . . .

The foregoing has illustrated the difficulty of setting objective and meaningful safety standards in a new industry carrying some risk. A not dissimilar problem exists in the drug and pharmaceutical industry. In both cases a recognition in advance of the possibility of harm, even at a low level of risk (frequency), and considerable difficulty in describing the risks exists and, as yet, there is no satisfactory way of agreeing on "what is reasonable" either within or outside the nuclear community.

I have quoted various statements which have referred to risk—in general they have shed little light on the problem and often revert to a circular argument. For example, in 1966, in part 100 of its regulations, "AEC has in effect defined undue risk by establishing general guidelines for evaluating the safety of reactor sites"—by implication, it is safe if it is sited.

It has been generally supposed that radioactive poisons are very much more hazardous them chemical ones (stated by McCullough [1] as a factor of 10^6-10^9). The basis for comparison is important; if we consider that quantity of "poison" which has a 50% chance of causing death, then we find the ratio by weight of chlorine to plutonium is nearer ten to one than a million to one. If, on the other hand, we consider, on a linear hypothesis, the dose which might give one chance in a thousand of death after many years, then we arrive at a lower figure for plutonium by a factor of 1000.

I am not here attempting to justify risk at any level nor to justify one risk by comparison with another of a different type. I seek rather to clarify the nature of the risk we run in an industrial society—a society in which the risk of accidental death decreases annually, and in which life expectancy annually increases. There is a small risk of serious accidents which we should endeavor to anticipate by paying greater attention to the less severe events which now more frequently occur.

We cannot achieve zero risk; we can reduce the likelihood of serious accidents by the effective application of relevant technical and managerial skills. The chapters which follow aim at promoting interest in the many varied aspects of safety technology.

References

1. McCullough, C. R., *et al.* The safety of nuclear reactors, Proceedings of International Conference on the Peaceful Uses of Atomic Energy, Geneva, 1955, p. 79, Vol. 13.

2. Hinton, Sir Christopher. The Future for Nuclear Power, Axel Ax, Son Johnson Lecture, Stockholm, March 15, 1975.
3. Parker, H. M., and Healy, J. W. Environmental effects of a major reactor disaster, Proceedings of International Conference on the Peaceful Uses of Atomic Energy, Geneva, 1955, p. 106, Vol. 13.
4. USAEC. WASH-750, Theoretical Possibilities and Consequences of Major Accidents in Large Nuclear Power Plants, March, 1957.
5. Marley, W. G., and Fry, T. M. Radiological hazards from an escape of fission products and the implications in power reactor location, Proceedings of International Conference on the Peaceful Uses of Atomic Energy, Geneva, 1955, Vol. 13, pp. 102-105, United Nations, New York, 1956.
6. Proceedings of IAEA/ISO Panel on the Safe Siting of Nuclear Reactors, Vienna, Oct. 31-Nov. 3, 1961, pp. 7-10.
7. Farmer, F. R. The evaluation of power reactor sites, Proceedings of CNEN Conference Problemi di Sicurezza degli Impianti Nucleari, VII Congresso Nucleare, Rome, June 11-17, 1962, pp. 39-45.
8. Beck, C. F. Engineering out the distance factor—A progress report on reactor site criteria, Annual Convention of the Federal Bar Association, Philadelphia, Pennsylvania, September 25, 1963.
9. Farmer, F. R. The growth of reactor safety criteria in the United Kingdom, Anglo-Spanish Nuclear Power Symposium, Madrid, November, 1964.
10. Hughes, H. A., and Horsley, R. M. Application of safety limitations against depressurisation to the Calder and Chapelcross reactors, BNES Symposium on the Safety of Magnox Reactors, London, November 11, 1964, pp. 198-203.

1
Introduction
H. M. Nicholson

Wartime research demonstrated both the explosive potential of the atomic bomb and the feasibility of power generation from controlled reactions in the atomic pile. This association of the military and peaceful facets of "atom splitting," together with the highlighting of the hazards of the radioactive fragments, gave rise to an acute concern for safety in the development of nuclear power. The result has been an unprecedented effort on safety studies and a low accident record unmatched in any other modern industry.

Over the last thirty years research and development has continued to resolve the technical problems which somewhat delayed the early promise of cheap, abundant atomic power. Today that promise is becoming a reality. In parallel with this technological growth, safety research has defined ever more precisely the associated hazards. As power stations have come into operation around the world, public awareness of the potential hazards of the reactor has increased and fear of possible accident consequences has generated vigorous reactions from those concerned with conservation of the environment. In the heat of public debate many of the conclusions have been distorted by arguments based on questionable interpretation or selection of the data.

One aim of this monograph is to put the nuclear hazard in perspective by an objective overall technical review of the field which will recognize the nature of the hazards, assess their gravity, and attempt to show that appropriate steps are being taken to ensure that the advantages to the community are commensurate with the inevitable risks.

The subject of reactor safety covers a wide field of science including engineering, physics, chemistry, biology, meteorology, geology, ecology, and mathematics, and during the last twenty years a vast literature has been generated. This monograph can be little more than an introduction to this expanding field of study. The readership envisaged is twofold. First, the technicians, those who are engaged in nuclear engineering: designers, constructors, and operators of nuclear stations, as well as those who would make a career in nuclear safety. Second, those (not necessarily scientists) who are carrying the responsibility for making decisions in the field of energy use and allocation or are concerned with environmental matters. It is to be hoped that the text will not prove to be too elementary for the first group nor too obscure for the second.

With such a wide field of technology to cover, there is a need to be selective in a monograph of this size. Accordingly, we have chosen not to deal with the safety or regulatory aspects of normal operation which are designed to protect the operators, the general public, and the environment from significant harm. The management of waste products, the development of safe practices in containing and transporting nuclear materials, and the means for preventing the occurrence of unplanned critical nuclear reactions in transport, storage, and chemical processing are all important and absorbing topics that have been studied nationally and internationally and continue to receive much attention. However, these specific topics are consciously omitted from this monograph in order to deal more fully with the main theme: that of reactor accidents and their consequences. The technical arguments will be concerned broadly with reactor accident conditions and will deal with both the arrangements necessary to prevent any dangerous diversion from normal operation and to ameliorate the consequences if such a diversion should occur. A recurring feature throughout the monograph will be the attempt to quantify the arguments. Since it is unrealistic to expect all risk to be eliminated, the concept will be promoted that the acceptable probability of a particular accident occurring should be inversely proportional to the severity of that accident. Both the probability and the consequences will be quantitatively assessed and a constant for the inverse proportionality will be suggested. It will be demonstrated that quantification is more readily achieved in some areas than others and that there is still considerable field for judgment.

The monograph falls into three main sections as follows:

(a) In Chapter 2 the nature of fission products is described and their physical and chemical properties which are

1. Introduction

significant in safety studies are highlighted. These properties, combined with their relative abundance and biological reactions, serve to identify the few nuclides important in safety studies. The generation, migration, and containment of these nuclides is discussed when relevant for accident conditions. Chapter 3 deals with the hazards to man and his environment which result from the uncontrolled release of fission products in accident conditions. This chapter will show how accidents may range over a very wide spectrum from those having small consequences, either in relation to people, the environment, or the economy, to those having very severe consequences in one or more of these aspects.

(b) In Chapter 4 a quantitative approach to reactor safety assessment is developed and a numerical criterion for acceptability of risk is proposed. Risk to both the individual and the community is considered and evaluated relative to the other risks of normal living. The influence of population distribution, meteorology, geology, and communications on reactor siting are discussed and quantified where possible. Chapter 5 is concerned with demonstrating the quantitative approach in the field where it has proved most effective, i.e., the reliability and safety of protective systems. The application of such techniques has also proven effective in deciding the strategy to be adopted in engineering emergency cooling systems—this is discussed in Chapter 6. Chapter 7 on the other hand explores the more difficult field of quantification of vessel integrity. Here experience and limited failure data provide the basis for an engineering judgment rather than a confident prediction; human judgment, preferably expressed in numerical terms, remains an important element in analysis. Experience over a number of years has shown that the discipline imposed by the pursuit of quantitative assessment in all areas has yielded a more ordered and comprehensive approach to safety studies. A major benefit is the aid which accrues in identifying those areas of study where a limited technical effort can be most profitably employed.

(c) Chapters 8 and 9 deal with the basic principles of analysis and assessment of reactor safety and then consider the specific safety problems of thermal and fast reactors in detail.

This monograph brings together contributions from a number of authors, all specialists in their various fields. All are members of the staff of the Safety and Reliability Directorate of the United Kingdom Atomic Energy Authority and they are identified at the head of each chapter. Thanks are due to all these authors and to the Chairman of the UKAEA for permission to publish this material.

2
Radioactivity and the Fission Products
F. Abbey

Introduction

   The fission or splitting of nuclei of the fuel atoms is the key event by which heat is produced in nuclear fission reactors, whether thermal or fast. Atoms of the fuel combine with an incident neutron to form unstable nuclei which then achieve stability by splitting into two or more fragments, generally of comparable size. A few additional neutrons are released at the same time, and these are necessary to the operation of the reactor since they initiate the further fissions which produce a continuous chain reaction.
   It is the fission products, that is the elements near the middle of the periodic table created by fission of the fuel, which by virtue of their radioactivity constitute the major and characteristic hazard of the fission reactor. The hazard to the operator from fission products retained within the reactor, and from a fraction of the neutrons and gamma rays generated in fission which escape from the reactor core, is easily reduced to acceptable levels by suitable shielding. The principal hazard is then from fission products released to the environment. The nature of this hazard and the consequences of the accidental release of substantial quantities of fission products are enlarged on in Chapter 3. This chapter is concerned with the general ideas of radioactivity, its effects, and with the amounts, properties, and behavior of the fission products formed in a reactor.
   The treatment in the chapter is given in general terms applicable to most reactor systems. However, in the USA

information on fission product species build-up, transport, and removal for LWRs is in course of codification, and standards providing quantitative guidance for these reactors are available or in preparation. Standard assumptions acceptable to the authorities for licensing purposes are also defined in regulations and regulatory guides. Where appropriate, this chapter makes reference to these documents.

Other radioactive species apart from fission products are also formed in fission reactors by the absorption of neutrons in nonfissionable atoms outside the fuel, but these so-called activation products usually represent a much smaller hazard than the fission products and are not discussed further.

Radioactivity and Its Effects

Radioactive Decay

Radioactivity is the phenomenon by which certain types of atoms or nuclides which are unstable achieve stability by rearrangements within the nucleus, accompanied by the emission of particles and radiation. It was first discovered in some of the heavy elements beyond lead in the periodic table. These were found to be naturally radioactive and the emanations were classified into three types, alpha particles, beta particles, and gamma rays. Experiments have shown that, in general, a single radioactive nuclide does not emit both alpha and beta particles, but that gamma rays may accompany either. Alpha particles have been shown to be identical with the nucleus of the helium atom, containing two protons and two neutrons, thus carrying two positive charges. This is a very stable unit and there is evidence to suggest that it exists as a subunit of more complicated nuclei. Beta particles are electrons which are supposed to be emitted as a result of the conversion of a neutron to a proton within a nucleus. While alpha particles are emitted with a definite energy, beta particles display a continuous spectrum of energies up to a maximum value. Conservation of energy is maintained in beta emission by the release of an additional particle—the neutrino—at the same time as the beta particle. This has no electrical charge and a small mass compared even with the electron, but it carries some of the energy of disintegration of the nucleus. Because it has very little interaction with matter the neutrino has no safety significance. The energy carried away by the combination of a beta particle and a neutrino or by an alpha particle will take one of several values characteristic of the parent nucleus and equal to or less than the total energy

2. Radioactivity and the Fission Products

of disintegration or decay (minus the recoil energy of the nucleus). Where the particle energy is less than the full decay energy, the daughter nucleus is left in an excited state and immediately emits a gamma ray accounting for the difference. Gamma rays are penetrating electromagnetic radiations similar to X rays but of higher energies.

Emission of either an alpha or beta particle is clearly an example of transmutation in that the atomic number and hence, the chemical identity of the atom is changed. The resulting daughter nucleus may itself be unstable and a radioactive series may then be formed ending when the chain is finally terminated by a stable nuclide. There are three natural radioactive series, referred to as the uranium, actinium, and thorium series and starting, respectively, with ^{238}U, ^{235}U, and ^{232}Th and ending with ^{206}Pb, ^{207}Pb, and ^{208}Pb.

Radioactive decay is a first-order process in that the rate at which atoms disintegrate is proportional to the number of atoms present, i.e.,

$$dN(t)/dt = -\lambda N(t) \tag{1}$$

where $N(t)$ is the number of atoms present at time t and λ is a constant known as the decay constant. Integrating this expression gives

$$N(t) = N_0 \exp(-\lambda t) \tag{2}$$

where N_0 is the initial value of $N(t)$. The time needed for N_0 to be reduced by a factor of 2 is given by

$$T_{1/2} = (\ln 2)/\lambda \tag{3}$$

It will be observed that $T_{1/2}$, which is known as the half-life, is independent of N_0 and is characteristic only of the nuclide concerned. It can be shown that the mean life of an atom of a radioactive nuclide is given by

$$T_m = 1/\lambda \tag{4}$$

Generalizing this treatment for the mth element in a radioactive series gives

$$dN_m/dt = -\lambda_m N_m + \lambda_{m-1} N_{m-1} \tag{5}$$

To determine the amounts of the various members of a series of n radioactive nuclides which are present at a given time, a set of n such equations must be simultaneously solved.

Radiation Effects

When alpha and beta particles and gamma rays impinge on materials, their energy is absorbed and may cause changes and

damage in the material. The most serious aspect of this, in
the context of safety, is the damage which may be caused to
living tissue, leading to sickness and even death of the individual
concerned. In general terms, high-energy particles
and radiation can cause damage of three kinds. The first effect
is transmutation of the constituent atoms of the material
into atoms of new materials which are foreign to their surroundings,
and may themselves be radioactive. The second effect
is displacement of atoms from their normal positions in
the structure of the material. Both of these effects involve
interaction of the incident particles or radiation with the
nuclei of the atoms of the target material and, therefore,
neutrons which carry no charge are particularly effective.
The third effect is by ionization, i.e., by removal of electrons
from atoms lying in the path of a charged particle and
the creation of ion pairs. Gamma rays, being electrically
neutral, are incapable of direct ionization but do cause ionization
indirectly by setting in motion charged particles with
which they collide. For alpha and beta particles and gamma
rays resulting from radioactive decay, ionization is the overriding
mechanism by which energy is transferred to tissue.
The majority of ion pairs created in this way recombine and
the final result is heating of the material; indeed it is this
mechanism by which the energy of fission, which appears largely
as kinetic energy of the fission fragments, is converted
into heat. However, the ion pairs are chemically reactive
and a proportion may take part in chemical reactions from which
biological effects in tissue may eventually spring.

Both alpha and beta particles are of low penetrating power
and can be stopped by relatively insignificant amounts of material.
However, whereas the relatively high mass and charge
of the alpha particles produce a linear path with a high specific
ionization per unit path length, the smaller mass and
charge of the beta particles lead to frequent scatters and a
nonlinear path with a much lower specific ionization per unit
path length. Gamma rays are much more penetrating than either
alpha or beta particles and may only be stopped by several
feet of heavy shielding or several hundred feet of air.

Fission Product Quantities
and Relative Importance

The fragments into which fissionable atoms are split are
not always the same, and after irradiation in a reactor, fuel
elements will contain up to a few percent of nearly 200 different
isotopes of nearly 40 different chemical elements
(atomic numbers 30-66). Figure 1 shows graphs of fission

2. Radioactivity and the Fission Products

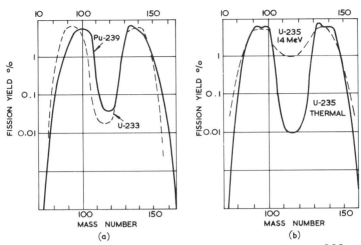

FIG.1 Fission yield versus mass number for (a) ^{233}U and ^{239}Pu and (b) 14 MeV and thermal ^{235}U.

yield (atoms formed per fission) against mass number for the common reactor fuels ^{235}U, ^{239}Pu, and ^{233}U, and some of the more important values are listed in Table I. Standard fission product yield tables for ^{235}U, ^{238}U, and ^{239}Pu at thermal, fission spectrum, and 14-MeV neutron energies are being prepared in the USA by ANSI [1]. It may be noted that nuclides whose mass numbers lie in the ranges of approximately 85-105 and 130-150 have a high yield. Many of the fission products are radioactive and decay by the emission of beta particles and gamma rays.

The precise calculation of the quantities of the various fission products present in a fuel element at any time during and after irradiation is a fairly complicated process, best carried out using a computer. It involves a preliminary calculation of the changes in the amounts of the fissionable nuclides such as ^{233}U, ^{235}U, ^{239}Pu, and ^{241}Pu which exist in the reactor as irradiation proceeds, then the calculation of the fission products formed, and finally, where appropriate, their subsequent conversion by radioactive decay and neutron absorption. The fission yields to be used may depend very slightly on the energy of the incident neutron. For an approximate calculation, however, one may ignore some of these complications (except for certain isotopes such as ^{106}Ru where the markedly different yields from ^{235}U and ^{239}Pu need to be taken into account, and ^{135}Xe which is a strong neutron absorber) and write

$$A = 8.4YM[1-\exp(-t/T_m)] \quad kCi \tag{6}$$

TABLE I

Some Fission Yields of Importance
(% Yield from Thermal Fission)

Isotope	Half-life	U-233	U-235	Pu-239
Kr-85	10.6 yr	0.58	0.293	0.127
Kr-83	Stable	1.17	0.554	0.29
Kr-84	Stable	1.95	1.00	0.47
Kr-86	Stable	3.27	2.02	0.76
Xe-133	5.27 days	—	6.62	6.91
Xe-135	9.2 hr	—	6.3	—
Xe-131	Stable	3.39	2.93	3.78
Xe-132	Stable	4.64	4.38	5.26
Xe-134	Stable	5.95	8.06	7.47
Xe-136	Stable	6.63	6.46	6.63
I-131	8.05 days	2.9	3.1	3.77
I-133	20.8 hr	—	6.9	5.2
I-135	6.7 hr	5.5	6.1	5.7
I-129	1.7×10^7 yr	—	0.8	—
Te-132	77 hr	4.4	4.7	5.1
Cs-135	2.6×10^6 yr	6.03	6.41	7.17
Cs-137	30 yr	6.58	6.15	6.63
Cs-133	Stable	5.78	6.59	6.91
Sr-89	50.5 days	5.86	4.79	1.71
Sr-90	28 yr	6.43	5.77	2.25
Ru-106	1.01 yr	0.24	0.38	4.57
Ba-140	12.8 days	5.4	6.35	5.4
Ce-144	280 days	4.5	6.0	3.79

where A is the fission product activity in kilocuries (one curie is that quantity of a radioactive nuclide giving 3.70×10^{10} disintegrations per second), Y is the percent fission yield, and M is the steady power in megawatts. Complications arising from short-lived precursors of the fission product of interest are ignored in this equation. When the radioactive mean life T_m is short compared to the irradiation time, the activity A reaches an equilibrium value,

$$A = 8.4YM \quad \text{kCi} \tag{7}$$

Equation (7) is useful for calculating the activity of nuclides such as ^{133}Xe and ^{131}I. When the radioactive mean life is very long compared to the irradiation time, the activity A increases approximately linearly with time,

2. Radioactivity and the Fission Products

$$A = 8.4 YMt/T_m \quad kCi \tag{8}$$

Important examples of nuclides showing this type of behavior are ^{90}Sr and ^{137}Cs.

If nuclear reactors are expected to operate efficiently, reliably, and safely, it must be possible to adequately inspect them and carry out maintenance and repair work to this end, and it must be possible to treat and store or dispose of any effluents. The formation of fission products in the reactor has important consequences for the reactor designer in all these areas. For example, some of the fission products such as ^{135}Xe are important neutron poisons and affect the reactivity of the core; the volume change consequent on fission product formation may cause fuel swelling and interaction with the fuel can, possibly accompanied by can corrosion; the decay heat of the fission products must be removed at all times, even when the reactor is shut down; and the reactor primary circuit has to be shielded against radiation from fission and activation products in the core and coolant. In the present context of safety in accident conditions, however, the aspect of fission products which is of principal concern is the possibility and effects of their release to the environment. Viewed in this light the problem of analyzing the behavior of the complex mixture which they represent can be considerably simplified. To reach the environment the fission products must escape from damaged fuel, from the reactor circuit, and from any surrounding containment. The potential release of a particular fission product element will, therefore, be dependent on the volatility and chemical reactivity of the product and its chemical compounds, and only relatively volatile materials are likely to be released in significant quantities. The effect of any materials which do escape will in turn depend on their radioactive half-lives and radiobiological properties. When all the relevant factors are taken into account usually only a few nuclides produced in high fission yield will be seen to merit consideration as important hazards to health. These include certain isotopes of krypton, xenon (generally referred to as the fission gases), iodine, tellurium, cesium (generally referred to as the volatiles), strontium, and ruthenium. Strontium has limited volatility as the element and is essentially involatile as the oxide, while the reverse is true for ruthenium which is involatile as the element but forms volatile oxide species. These two elements require special consideration, therefore, in that their behavior is likely to be influenced by the oxygen potential of their surroundings. Table II [2] gives the activities of some of the relevant nuclides, of which ^{131}I and ^{137}Cs are especially important. The release of ^{131}I, in particular, has

TABLE II

Activities of Some Fission Products in Uranium Fuel

Isotope	Half-life	Activity[a] (kCi/MW)	Isotope	Half-life	Activity[a] (kCi/MW)
I-131	8 days	30.4	Cs-137	33 yr	3.3
I-132	2.3 hr	37.6	Sr-90	28 yr	2.4
I-133	21 hr	55.5	Sr-89	54 days	24.2
I-134	52 mon	66.7	Ru-106	1.0 yr	15.9
I-135	6.7 hr	52.2	Ru-103	40 days	40.7
Te-132	77 hr	37.6	Ba-140	12.8 days	50.1

Time since shut-down (sec)	0	10^2	10^3	10^4	10^5
Total gases (Ci-MeV beta/MW) ($\times 10^5$)	4.40	1.61	0.565	0.188	0.099
Total gases (Ci-MeV gamma/MW) ($\times 10^5$)	5.70	2.89	1.18	0.545	0.081
Total volatiles (Ci-MeV beta/MW) ($\times 10^5$)	10.35	4.75	2.69	1.25	0.545
Total volatiles (Ci-MeV gamma/MW) ($\times 10^5$)	17.25	11.8	6.98	3.06	1.47

[a] Irradiation time = 1000 days. Cooling time after shut-down = NIL.

frequently been used in the past as a standard measure of the severity of a reactor accident because its property of concentrating in the thyroid gland following ingestion makes it the greatest potential hazard to the general public (see Chapter 3).

Although the behavior of the fission products can be understood in terms of their chemical and physical properties, it needs to be borne in mind that because of the low masses and concentrations of materials involved, the behavior may sometimes appear to depart from that of the same materials in bulk. For instance, surface effects and reactions with small amounts of impurities may become important. Examples of such effects are given later and one consequence is that studies of the behavior of radioactive nuclides must take account of the existence of stable isotopes of the same element, e.g., the mass of ^{131}I reaches an equilibrium value of about

2. Radioactivity and the Fission Products

TABLE III

Mass of Fission Products

Element	mg/MW day	Element	mg/MW day	Element	mg/MW day
Ge	0.011	Ru	65.4	Ba	38.6
As	0.003	Rh	17.1	La	39.8
Se	1.20	Pd	33.4	Ce	86
Br	0.36	Ag	2.7	Pr	37
Kr	10.4	Cd	1.67	Nd	140.6
Rb	10.2	In	0.08	Pm	8.86
Sr	28.2	Sn	0.97	Sm	27.2
Y	15.2	Sb	0.53	Eu	3.48
Zr	119.6	Te	15.7	Gd	0.036
Nb	0.33	I	5.86	Tb	1.67
Mo	107	Xe	149	Dy	0.005
Tc	27.4	Cs	90.4		

0.3 g/MW in the fuel, but this is eventually exceeded by the combined mass of stable ^{127}I and very long-lived ^{129}I which accumulates at the rate of 2.5 g/y MW. The total mass of iodine released in an accident may affect *inter alia* the proportion removed by plate-out on containment surfaces or by filter systems. The masses of fission product elements generated are listed in Table III.

Aside from the fact that their release is the end product of an accident, the fission products may also be important in influencing the course and mechanism of the accident. For example, the fraction of the largely stable isotopes of krypton and xenon which escape from the fuel material in normal reactor operation will pressurize the fuel can and may be instrumental in causing the can to fail in an accident. The total yield of krypton and xenon generated is equivalent to about 25 cm^3 NTP/MW day.

The Chemical Form of Fission Products and Their Behavior in Normal Reactor Operation

The form and disposition of the fission products in normal reactor operation is important to studies of their behavior in accident conditions because it represents the starting point from which these studies must begin. If there are substantial inventories of fission products outside the fuel material, and even outside the fuel cans if the reactor has been run with some elements with failed cans, then the safety of the reactor may depend on the behavior of these inventories in the accident. The USAEC's Interim Licensing Policy on "as low as practicable" for gaseous radioactive releases from light water cooled nuclear power reactors [3] specifies certain standard assumptions regarding the disposition of fission product iodine outside the fuel cans in normal operation of these reactors.

Release from the Fuel

In reactors with metallic fuels, for instance the Magnox stations which have uranium metal fuel rods clad in magnesium alloy cans, the principal modes of release of fission products from the fuel material are related to the fission process itself. The considerable kinetic energy of recoil initially possessed by a fission fragment results in a proportion of those formed within range of a free surface being released. The majority of these recoil atoms are normally reembedded in the can but some will be retained in any fuel/can interspace. An associated phenomenon known as "knock-out" occurs when a fission fragment displaces atoms of both fuel material and previously formed fission products from the fuel surface. The range of fission fragments, like that of alpha particles, is very small and it is clear that "recoil" and "knock-out" will normally release only a small fraction of the fission products formed in the fuel material. Such release is usually only of interest as the basis of methods for detecting and locating fuel elements with failed cans.

For modern power reactors fueled with porous ceramic materials operating at higher temperatures than metallic fuels, increased mobility of the fission products within the fuel material subsequent to their formation, either by diffusion or by recrystallization of the fuel material, leads to releases many times larger than those due to recoil and knock-out. Fission products which are soluble in the fuel material will remain in solid solution with consequent modification of the properties of the fuel, but the majority are likely to be highly insoluble in the fuel and will precipitate at the first favorable opportunity. They will then migrate to form bubbles of gas or condensed inclusion phases. In the hotter regions of the fuel the differential vapor pressure of the fuel material on either side of a bubble causes the bubble to migrate up the temperature gradient, leading to the formation or enlargement of central voids and the release of the bubble contents. In the cooler regions of the fuel, bubbles and inclusions agglomerate principally at grain boundaries and fission product release then occurs when the bubble concentration at

2. Radioactivity and the Fission Products

the boundary is sufficient for the bubbles to interlink and create a path to an open pore surface. The formation of paths of this kind may depend on fuel/clad stresses and once created they may remain open or may resinter until sufficient gas reaccumulates, according to circumstances. For fission products other than the inert gases krypton and xenon, participation in these mechanisms will be affected by chemical interactions. In the most common oxide systems, i.e., UO_2 or $(U, Pu)O_2$, the influence of the oxygen itself is of primary importance and fission products can exist in varying states of oxidation depending on the oxygen potential of the fuel material and the temperature. The oxygen content of the fuel material is defined by the initial oxygen-to-metal ratio, though there may be some redistribution of the oxygen under the influence of the temperature gradients and compositional changes accompanying irradiation. Owing to an imbalance between the oxygen demands of the fission products as compared with the fissile atoms destroyed, progressively more oxidizing conditions are established as irradiation proceeds [4]. The stable oxide-forming elements (Sr, Ba, Y, La, the rare earths, and Zr) will form oxides at all conditions of practical importance and will exist as solid solutions in the fuel material or separate oxide phases. When the oxygen potential is sufficiently low, other fission products will remain as the elements and are likely to behave as gases when sufficiently volatile. Cs, Rb, Te, I, and Br fall into this category. Some complications are possible, in principle, due to the formation of compounds between the fission products themselves, e.g., CsI, but no modification of behavior has been found that can be attributed to this type of reaction. At higher oxygen potentials oxidation of fuel material or cladding can occur and oxygen progressively dissolves in the cesium. At still higher oxidation potentials oxidation of Mo can occur and at very high oxygen potentials, that can only be achieved by using initially substantially hyperstoichiometric fuel, additional fission products become volatile (Mo, Te, Ru, and Rh), due to the formation of oxide species. The interactions in carbide fuels are similarly complex.

The extent of release of volatile fission products from ceramic fuel materials by the processes described depend on the type of fuel and its method of manufacture, and on the operating conditions. These in turn are functions of the type of reactor. For UO_2 at temperatures up to 1600°C it is possible to think of the accumulation of fission products in bubbles at grain boundaries as governed by simple diffusion theory, and by modifying this by some factor representing the extent of development of interconnecting paths between

the fuel/can interspace and grain boundaries the release to
the interspace can be calculated. At temperatures above 1600°C
diffusion theory can be expected to break down in oxide fuels
because recrystallization of the fuel occurs: in the temperature range 1600-1800°C there is equiaxed grain growth and from
1800°C to the melting point of 2800°C large columnar grains
are formed and there is gross bubble movement leading, as seen,
to the growth of central voids. Release of the inert-gas
fission products and iodine increases with increasing temperature throughout this range from very small values at 700°C or
so, where recoil and knock-out are the principal contributions,
to about 30% in the region of equiaxed grain growth and 100%
in the region of columnar grain growth. Below 1600°C the release is dependent on half-life and irradiation time but where
grain growth occurs the equilibrium release is reached quickly
and is, therefore, independent of half-life for the principal
nuclides of interest in safety studies, notably ^{131}I [5]. In
practice, release from regions above 1800°C usually dominates
the overall release of inert gases and iodines from oxide
fuels in the more highly rated LWRs and LMFBRs but for the gas-cooled reactors, fuel center temperatures are unlikely to
exceed 1600°C and releases are generally smaller. Substantial
releases of cesium to the fuel/can interspace can also be expected in reactors fueled with oxide fuel. Recommendations
regarding fuel plenum gas activity for LWRs in the USA are
being prepared by ANS [6].

It does not necessarily follow that fission products which
are released from the fuel material of a fuel element in normal
operation will exert a high vapor pressure on the fuel can.
Chemical interactions between the different fission products
or with the can, and the relatively low temperature of the can
surface, may result in much of the fission products being removed from the vapor phase. The release from the fuel element
to the reactor circuit which occurs when the can fails in normal operation may reflect this, although access of the reactor
coolant to the interior of the can may disturb the situation
from that in the intact fuel element. For example, it is possible to compare the releases of inert gases, for which there
can be no retention in the fuel can, and the release of ^{131}I;
from such comparisons and from other evidence it has been concluded in a number of cases that only from 1% to 10% of the
^{131}I released from the oxide fuel material of Zircaloy-clad
water reactor fuel elements subsequently escapes to the reactor coolant [7]. In other instances the disturbance of the
situation from that in an intact fuel element which occurs
when an element fails in normal operation may extend to changes
in the fuel material itself. For instance, in the AGR the

2. Radioactivity and the Fission Products

UO_2 fuel may be subject to oxidation by the carbon dioxide coolant, in the oxide-fueled LMFBR there may similarly be fuel/coolant interaction to form a mixed fuel/sodium oxide, and in the helium-cooled HTGCR oxide fuel may be converted to carbide by reaction with adjacent carbon. These interactions can be expected to affect the release of fission products. It is known that where the fuel oxidizes there will be a general increase in the release of the fission gases and volatiles.

Whereas most reactors use metal-clad fuel, in the HTGCR the fuel consists of small particles coated with layers of pyrolytic carbon and silicon carbide. Early experiments on particles coated only with pyrolytic carbon showed that strontium, and possibly cesium if the irradiation time was long enough, were capable of diffusing through the pyrolytic carbon. A coating consisting of a thin layer of silicon carbide between layers of pyrolytic carbon was developed to reduce the release of the metallic elements. A very small fraction (which is likely to be of the order of 1 in 10,000) of the particles in a fuel element can be expected to have defects in one or more of these layers owing to manufacturing imperfections or failures in service, and there will be a consequent unavoidable release of fission products. However, the structure of the fuel element is generally such that the particles are molded into a matrix of graphitic material which is then enclosed within a further region of fuel-free graphite, and these regions will retain at least a large part of the cesium and strontium.

The overall picture which emerges, therefore, is that fuel elements which fail in normal operation release fission products into the reactor circuit, and that the amounts released are frequently less than has escaped from the fuel material itself. The precise proportions are, however, a function of the type of fuel, the operating conditions, the geometry of the failure, and the extent of coolant ingress.

Behavior outside the Core

Once released to the reactor circuit a variety of fates await the fission products: they may escape from the circuit by leakage, they may be removed by various kinds of coolant clean-up systems, they may plate out on the surface of components of the circuit, or finally they may remain distributed in the coolant. Owing to its chemical reactivity iodine, in particular, can exist in a variety of forms in reactor circuits depending on circumstances. In thermal reactors three types of iodine are generally distinguished: "inorganic" iodine which is largely elemental iodine and its reaction products with water, iodine attached to particulate matter, and "organic" iodine. The latter is a vapor form of low chemical reactivity found in at least small proportions in almost every study of iodine behavior. Because it does not easily deposit on surfaces or into water solutions this material often determines the overall efficiency with which fission-product iodine can be contained and considerable efforts have, therefore, been devoted to its identification. These have shown that it generally consists of mixtures of alkyl iodides with methyl iodide predominating. The formation of these organic species clearly depends on the presence of small amounts of carbonaceous impurities and is illustrative of the general feature of fission-product studies referred to earlier, that because of the low masses and concentrations of materials concerned their behavior may well appear to depart from that of the same material in bulk. Methyl iodide is the principal form of iodine in the carbon dioxide coolant of the AGR, where it is unstable and has a lifetime varying between a few seconds and a few minutes depending on reactor operating conditions and temperatures. In the helium coolant of the HTGCR and in water reactors, the iodine probably exists mainly in the inorganic form. In the LMFBR, iodine in whatever form it is released from the fuel can be expected to react with the sodium coolant to form sodium iodide (NaI). Other fission products released to the circuits of various reactor systems generally display much less versatility than iodine and probably exist either as molecular elemental species (notably the inert gases and cesium) or as particulates in elemental or oxide form. There is again no evidence of compounds formed between the different fission products, e.g., CsI. In gas-cooled reactors all but a very small fraction of the fission products released from the core in normal operation, apart from the inert gases, plate out on circuit components, and particularly on the heat exchangers which offer a large surface area at relatively low temperature. The situation in other reactors is not so well defined at the present time.

Fission Product Behavior
in Accident Conditions

The release of fission products under accident conditions will clearly depend both on the reactor type and on the severity of the accident. The accidents which normally receive the most attention from this point of view are, for obvious reasons, those in which the reactor circuit is breached or those involving the handling of fuel outside the reactor

2. Radioactivity and the Fission Products

circuit. If the arrangements for shutting down the reactor and cooling the fuel function correctly in accidents of the first kind, there may be no consequent fuel can failures. This is the intended course of events, for example, in depressurization of the AGR or HTGCR and in an SGHWR or LWR loss-of-coolant accident. Where fuel can failures are successfully avoided, it is the fission product inventory of the coolant and the magnitude and stability of fission product deposits on circuit components which are important. By suitable restrictions on the amount of failed fuel permitted in the reactor in normal operation, the potential release of fission products to the environment can be kept to very low values, though there is some uncertainty at the present time about the stability of surface deposits in some accidents. In more severe accidents, however, additional release of fission products from fuel damaged in the accident may occur. Standard assumptions regarding the release of fission products in accidents for LWRs to be licensed in the USA are being specified by the USAEC. The annex to Appendix D of 10 CFR Part 50 specifies assumptions appropriate to environmental reports and a series of regulatory guides specifies assumptions appropriate to safety evaluations.

Release from the Fuel

The primary factor controlling the extent and nature of fission product releases from fuel damaged in an accident is the temperature attained by the fuel elements. Releases can vary widely and only limited generalizations can be made. For present purposes it is intended to distinguish two main categories of accident—those in which fuel element cans fail but do not melt, and those in which there is can and even fuel material melting.

In an AGR depressurization accident or in a loss-of-coolant accident on a water reactor there is a small probability, dependent on the efficiency of cooling achieved, that in parts of the core temperature equalization between the fuel material and the can plus decay heating might raise can temperatures to a point at which failure by internal gas pressure would occur. Since temperature equalization also reduces peak fuel temperatures, there will be a negligible additional release of fission products from the fuel material during this period and only gas-borne fission products in the fuel/can interspace will be released when the can fails. Laboratory experiments in which preirradiated fuel elements have been heated in a stream of carbon dioxide or steam and then punctured by a laser beam have confirmed that only the inert gases, iodine and

TABLE IV

Can	Gas composition	Irradiation (MWD/ton)	Release conditions
Zr	Sealed can	1300	Can puncture at 900°C
Zr	Sealed can	850	Can puncture at 1050°C
Stainless steel	CO_2/50% CO	1000	Heated 1900°C
Stainless steel	CO_2/5% CO	2000	1700-1800°C
Stainless steel	CO_2/50% air	100	1700-1800°C
None	He	trace	1980°C
None	CO_2	11,000	2800°C (molten)
None	Air	11,000	2800°C (molten)

[a] Release expressed as a percentage of the activity of the isotope present in irradiated fuel.

cesium, are likely to be released in significant quantities, together with some instances with very small amounts of barium and strontium. Comparisons of inert gas and ^{131}I releases suggest that, as in normal operation, there will be significant retention of ^{131}I (70-90%) by the can and even greater retention of ^{137}Cs. Some typical results are illustrated in Table IV [2,8].

Even more improbable but more severe accidents can be postulated in which fuel elements are sufficiently starved of coolant, or in which a nuclear transient leads to overheating and there is clad melting with exposure of the fuel material to the coolant. In the more extreme cases there may even be melting of the fuel material itself. In a fuel handling accident these events might occur in air. Relevant experiments have been carried out in the laboratory with small irradiated specimens and have shown substantial releases of the fission gases and volatiles, increasing in magnitude with increasing fuel temperature. As is to be expected, in oxidizing atmospheres there are large releases of ruthenium and small but

2. Radioactivity and the Fission Products

Measured Values of Release from Uranium Dioxide Fuel[a]

Xe,Kr	Isotope					Other activity released
	I	Te	Cs	Ru	Sr	
5	0.35	—	0.27	—	0.0031	Ba, 0.0011
2	0.20	—	0.037	—	—	Ba, 0.00032
20	11	0.6	32	—	0.002	—
—	44-70	16	4-11	—	0.2	—
—	25.47	—	10-33	25	0.03-0.3	Ba, 0.1
79.0	53.0	74.0	84.0	6.0	15.0	Ba, 57.0
99.9	99.9	99.0	96.6	79.1	0.6	Ba, 2.9 Rare earths, 2.3
100.0	99.8	93.3	97.4	92.5	0.4	Ba, 1.8 Rare earths, 3.9

measurable releases of strontium and barium, whereas in oxygen-free atmospheres ruthenium release is much lower and the release of strontium and barium increases. Typical results are again illustrated in Table IV. A recent review of these data [2] suggests that whatever the overall fission product release in an accident it can be assumed, as a convenient generalization, that the fission gases, the volatiles, and ruthenium will be released in the same proportions in which they existed in the fuel immediately prior to the accident and that the release of strontium and the rare earths will be one hundredth of that of the gases and volatiles expressed as a percentage of the fuel inventory. In particular instances this generalization may be quite pessimistic as regards ruthenium release. Recommendations regarding fission product release values during accidental fuel melting for LWRs in the USA are being prepared by ANS [6].

In principle, higher releases of the normally nonvolatile fission products, and indeed of the fuel material itself, might be obtained if an accident could cause the fuel to reach

extremely high temperatures, in the region of 3000-4000°C.
This is very difficult to achieve even in a very severe accident; molten fuel will slump and share its heat with reactor structural materials and will lose heat through internal convection to surfaces from which heat can be radiated. The possibility that very large power transients could lead to high fuel temperatures is considered for all reactors and the likelihood of these accidents is reduced to a sufficiently low level by design of the reactor, its core and control rod configuration, and by diverse means of controlling power.

This discussion of fission product release from the fuel in accident conditions has been conducted essentially in terms of the conventional modern power reactor with oxide fuel clad in stainless steel or Zircaloy cans. Special considerations apply in the case of metal-fueled reactors and the HTGCR, and these are summarized in the review referred to [2].

Behavior outside the Core

Because of the circuitous route that fission products must follow after release from the fuel to escape to the environment, a large proportion will certainly be removed on the way by natural processes. Escape may be further reduced by engineered safeguards deliberately designed to this end. In both cases the process is one of mass transfer from the gas phase to a liquid or a solid surface, followed if necessary by chemical reaction so that a volatile fission product does not reevaporate. In principle, this order could be reversed and chemical reaction in the gas phase could be followed by deposition, but this approach has not found practical application in reactor safeguards.

Within the reactor circuit itself many of the less volatile fission products may condense within a short distance of the overheated fuel but, according to circumstances, a fine fume may persist incorporating (again according to circumstances) some of the fission products and vaporized fuel and canning. Some iodine may be absorbed on this particulate but some will exist as "inorganic" molecular species and methyl iodide. In the experiments on fission product release from the fuel referred to in the previous section, the fraction of the released iodine which was in the form of methyl iodide was generally between 3% and 10%. The fission gases will, of course, remain gas-borne. In an accident in which the reactor circuit is not breached, there will be further long-term removal of fission products by the same mechanisms of plate-out and coolant cleanup which operate in normal reactor operation. In the particular case of the LMFBR, where the damaged fuel stays well

submerged in a deep pool of sodium, it is unlikely that more
than a small portion of the fission products, apart from the
inert gases, will even penetrate to the cover gas.

Where the reactor circuit is breached, however, substantial quantities of fission products may well escape from the circuit to the reactor containment or auxiliary building. Further removal of fission products is then likely within the containment by continued radioactive decay and by deposition onto the containment walls or possibly onto coolant particles in the containment atmosphere. Deposition will be governed by a deposition velocity (mass-transfer coefficient) V_g according to the relation

$$V\, dc/dt = -AV_g C \qquad (9)$$

where C is the gas-borne concentration of the fission product, V the volume of the containment, and A the surface area available for deposition. On integration this gives

$$C = C_0 \exp[-AV_g t/V] \qquad (10)$$

where C_0 is the initial value of C. The deposition half-life is given by

$$T_{1/2} = [\ln 2/(AV_g/V)] \qquad (11)$$

The value of V_g for iodine depends to some extent on the adsorptive properties of the containment surfaces, which may vary with manufacture and conditions of use. However, several series of experiments have shown that the value is likely to be in the region of 10^{-3} m/sec for metal surfaces and perhaps half of this for a containment where concrete surfaces predominate [9]. Very steamy, wet conditions make very little difference to the value of V_g. Rather similar behavior is to be expected of other fission products which exist as fine particulate. Values of V_g of this magnitude lead to deposition half-lives of a few hours for typical large containments.

In the case of iodine there is a limit to the reduction in the gas-borne concentration resulting from deposition set by desorption processes and by the presence of methyl iodide. Methyl iodide is removed much more slowly at temperatures likely to be encountered in containments, and any reduction in the gas-borne concentration by natural process can probably be ignored in the short term.

Some fission products can be expected to be taken up by water dripping from wet surfaces following a water reactor loss of coolant accident, but since the drop size is unlikely to be the optimum and the quantity of water is limited, it is improbable that fission product removal by this route will be comparable in effectiveness with deposition to surfaces. In an LMFBR accident involving a sodium fire, fission products can similarly be expected to be taken up by the sodium oxide aerosol produced by the fire and to suffer the same fate as the aerosol. Fission products present in the sodium which is burned will be partly carried away by the aerosol and partly retained by fire residues.

Although natural processes of the kind described may be capable of removing a large part of the fission products released in an accident, it would clearly be unwise to depend to too great an extent on such processes and it is customary to provide engineered safeguards to guarantee adequate fission product removal. Some of the forms of engineered safeguards employed are described in subsequent sections.

Filter Systems

The application of filter systems for gas cleaning has a long history in the chemical and related industries. The specification and design of such systems is primarily determined by the required filtration efficiency, the maximum quantity of gas-borne species with which the filter can be loaded while continuing to function efficiently, and the pressure drop available for impelling the gas through the filter. From earlier discussions it will be clear that reactor filter systems will usually have two distinguishing requirements: (a) the removal of both molecular species and particles of a wide range of sizes, and (b) high removal efficiencies from gaseous atmospheres which are relatively lightly contaminated in mass terms. They may also have to function satisfactorily at high temperatures and high humidities. These requirements are met in most cases by a system in which the essential elements are a high-efficiency ("absolute") particulate filter followed by a bed of activated carbon (or other material with a highly developed surface area) to remove molecular species by adsorption [10].

The so-called "absolute" particulate filters consist of an assembly of fibers in depth and are capable of very high removal efficiencies, often better than 99.95%, for particles down to very small (submicron) sizes. The dimensions of the fibers of which the filter is formed, and of the pores between, are generally greater than those of the particles to be removed and it is clear that the filter does not simply act as a sieve. Actual mechanisms of particle arrest include direct interception, inertial impaction, diffusion, and electrostatic attraction [11]. The most attractive filter materials are those which can be formed into a lap or sheet which can be pleated around spacers so that a large surface area is presented to

2. Radioactivity and the Fission Products

the gas stream in a small volume. Materials which have been used include cotton-asbestos mixture, cellulose-asbestos paper and, more recently, paper made from fine glass fibers. The resistance to high temperature and chemical attack of the latter can be particularly useful.

The purpose of the activated carbon element of the reactor filter system is essentially the removal of molecular forms of iodine. Of these, elemental iodine is so readily trapped on most types of activated carbon that it can be assumed that if methyl iodide can be retained, the removal of elemental iodine will automatically be assured. (This statement may need qualification for very low gas-borne concentrations.) The retention of the adsorbed activity must be guaranteed for long periods of time while the gas flow continues, and this requires that the activity be chemisorbed and not merely delayed by physical adsorption. In the UK, interest has centered on filter systems required to operate under conditions of temperature and pressure not far removed from atmospheric [12], though there has been some work on high-pressure and high-temperature operation in carbon dioxide in connection with coolant filters for AGR. It was found that when operating below a critical loading limit the logarithm of the decontamination factor achieved for methyl iodide by a bed of a particular activated carbon is directly proportional to the stay time of the carrier gas in the bed; where the decontamination factor = 100/(percent penetration of methyl iodide through the bed), and stay time is defined as the total volume of the bed divided by the volume flowrate of the carrier gas. This can be taken to imply a first-order rate of reaction between the carbon surface and the methyl iodide and identifies the quantity K (sec^{-1}) = log (decontamination factor)/stay time as a useful index of carbon performance. Filter systems are generally expected to handle flows of the order of 300 m^3/min or larger. Thus, if the carbon beds are to remain of reasonable size, stay time must be of the order of a fraction of a second, and this in turn implies K values in excess of one if decontamination factors comparable with those of "absolute" particle filters are to be obtained. It will usually be found that a bed sized to give an acceptable decontamination factor will provide more than adequate loading capacity for methyl iodide.

Investigation also revealed that the performance of activated carbon beds against methyl iodide is very adversely affected by the presence of water in the carbon and hence in the carrier gas. The water presumably reduces the carbon surface area available to methyl iodide. If the carrier gas approaches saturation with water, as it may well do in the case of a water reactor accident, acceptable decontamination factors cannot be achieved at any reasonable stay time. An improved carbon performance is obtained in humid conditions by impregnating the carbon with chemicals to increase the reactivity of the surface to methyl iodide. Two types of impregnant have been developed for application under UK conditions, the first is a series of heterocyclic amines of which the best is triethylenediamine (TEDA) and the second a series of inorganic iodides such as potassium iodide. Carbon impregnated with the recommended quantities of either of these impregnants will perform almost as well against methyl iodide at 100% carrier gas relative humidity as will unimpregnated material in dry gas. TEDA requires slightly smaller stay times and has the higher loading limit, but being a strong base its performance is adversely affected by acidic impurities, including carbon dioxide, in the gas stream.

In the USA, interest has extended to filters required to operate under the more extreme conditions of temperature, high humidity (steam) at high pressure, and high radiation levels which may exist in the containment of a LWR following a major loss-of-coolant accident. After exposure to steam for several hours the amine impregnated carbons may show a significant deterioration in performance due to volatilization of the impregnant, but iodide impregnated carbons are not so affected and are capable of good performance under these conditions. On the other hand, TEDA impregnated carbons may have advantages in resisting the effects of high levels of radiation.

Containment Sprays

A disadvantage of filter systems for the removal of fission products from large containment volumes is that the flowrates which the systems can accept are generally relatively small by comparison with the containment volume. Some time will elapse, therefore, before the entire containment atmosphere can be treated. An alternative engineered safeguard which does not suffer from this disadvantage is the containment spray system [13] which drenches the whole of the containment volume with a fine spray of water droplets from the instant of initiation. Such systems may perform the double task of condensing steam released in the accident and removing fission products.

The scrubbing efficiency of a spray system will depend on the number of spray drops per unit volume of the containment atmosphere (N) and on the size (diameter d*) of the drops, of

*In this simplified treatment, d is assumed to be the same for all drops.

2. Radioactivity and the Fission Products

which the former is a function of the mass of water sprayed per unit floor area of the containment. Transfer of fission products in molecular form to the drops is by diffusion and depends on the deposition velocity, and a formula analogous to Eq.(9) may be written

$$V \, dc/dt = VN\pi d^2 V_g \tag{12}$$

where $VN\pi d^2$ is the total surface area of spray droplets. Then by integration

$$C = C_0 \exp(-N\pi d^2 V_g t) = C_0 \exp(-\lambda_m t) \tag{13}$$

where m has been termed the "washout coefficient." Transfer of fission products in particulate form to spray drops occurs by a quite different mechanism. The diffusion coefficients of even quite small particles are several orders of magnitude lower than those of molecular species. Hence, diffusion is no longer important and the principal mechanism of uptake is by impaction; as a water drop traverses the cloud of particles it sweeps out a certain small volume. Not all the particles in the volume swept by a drop are captured; however, some are carried away by the air displaced by the falling drop. Thus, for particulate species the "washout coefficient" becomes

$$\lambda_p = (N\pi d^2/4) V_t E$$

where V_t is now the terminal velocity of the spray drops and E is an efficiency of impaction.

For drops in the usual size range encountered in containment sprays, i.e., 100-1000 μm, V_g is in the region 0.2 m/sec and V_t varies between 0.25 and 4 m/sec depending on drop size. The value of E depends sensitively on the size of the particle, becoming very small for submicron particles. It can be shown on these grounds, and has been experimentally confirmed, that removal of soluble molecular species such as "inorganic" iodine can be expected within a few minutes by practical containment spray systems. If there is concern about reevaporation of elemental iodine from drops as they become saturated, this can be overcome by adding to the spray solution various chemicals which are capable of reacting with and fixing the iodine [14]. Additives of this kind favored by the industry are sodium hydroxide and sodium thiosulphate. Owing to its low solubility in water or any of the preferred spray solutions, it is generally assumed that methyl iodide will not be well removed by sprays. As well as removal of "inorganic" iodine, sprays will also accomplish the rapid removal of the larger particles but because of the influence of the efficiency of impaction E it is generally assumed that submicron particles will not be well removed. It may be, therefore, that the overall efficiency

of sprays for the removal of those fission products which may
be expected to occur in particulate form in containment systems,
notably cesium, strontium, and ruthenium as well as iodine adsorbed on particles of other materials, may be rather low.

Passive Systems

A criticism which can be leveled at both filter systems
and containment sprays as engineered safeguards is that they
are both "active" systems in the sense that plant items such
as fans, pumps, etc., are required to operate correctly for
the system to achieve its purpose of removing fission products.
In contradistinction are such "passive" systems as the ice
condenser and suppression pond containments which do not involve working plant. In both of these the principal aim is
to achieve rapid condensation of steam released in a water
reactor loss-of-coolant accident by directing the steam either
through a compartment filled with ice or through a pipe dipping into a pond of water, but both systems also clearly provide an opportunity for fission product uptake. The disadvantage of such systems, which is the obverse of their advantage
of reliability of operation, is that they are essentially
"once through," i.e., with no working plant the containment
atmosphere cannot be recirculated through the system when the
initial discharge of steam has terminated, and this may restrict the levels down to which fission products can be removed
in practice. Theoretical analysis of fission product removal
of the kind which can be undertaken for sprays is difficult,
particularly for suppression ponds, because of uncertainties
in the hydrodynamics of the system. However, experiments have
shown that in both suppression ponds and ice condenser systems
fission product uptake can be substantial. For instance, experiments with suppression ponds in the UK [15] have shown that
fission product removal is dependent on the proportion of air
which is mixed with the steam passing through the pond but is
never likely to be less than 90% for a practical system and
may be substantially more.

References

1. ANSI, N.217, ANS Working Group 5.2, Standard Fission
   Product Yields for ^{235}U, ^{238}U, and ^{239}Pu at Thermal, Fission Spectrum, and 14-MeV Neutron Energies (in preparation).
2. Beattie, J. R., and Bryant, P. M., UKAEA Rep. AHSB(S)R135
   (1970).

2. Radioactivity and the Fission Products

3. USAEC Regulatory Guide 1.42 (Revision 1), Interim Licensing Policy on as Low as Practicable for Gaseous Radioiodine Releases from Light Water Cooled Nuclear Power Reactors (March 1974).
4. Findlay, J. R., Paper to IAEA Panel Meeting on Behaviour and Chemical State of Fission Products in Irradiated Ceramic Fuel, IAEA, Vienna, 1974.
5. Allen, J., *J. Brit. Nucl. Energy Soc.* 6, No. 2, 127 (1967).
6. ANSI, N.218, ANS Working Group 5.4, Fuel Plenum Gas Activity (in preparation).
7. Abbey, F., Paper to OECD/CSNI Specialist Meeting on the Safety of Water Reactor Fuel Elements, Saclay, October 22-24, 1973, CEA, Saclay, France, 1973.
8. Browning, W. E., Miller, C. E., and Shields, R. P., *Nucl. Sci. Eng.* 18, 151 (1964).
9. Chamberlain, A. C. et al., *J. Nucl. Energy Pt A/B*, 17, 519 (1963).
10. USAEC Regulatory Guide 1.52, Design, Testing and Maintenance Criteria for Atmosphere Clean-up System Air Filtration and Adsorption Units of Light Water Cooled Nuclear Power Plants (June 1973).
11. IAEA. Techniques for Controlling Air Pollution from the Operation of Nuclear Facilities, Safety Series No. 17, Vienna, 1966.
12. Collins, D. A., Taylor, L. R., and Taylor, R., UKAEA Rep. TRG 1300(W) (1967).
13. Symposium on Reactor Containment Spray Technology, *Nucl. Tech.* 10, No. 4, 405 (1971).
14. ANSI, N.581, ANS Working Group 56.5, PWR and BWR Containment Spray System Design (in preparation).
15. Hillary, J. J., Taylor, J. C., Abbey, F., and Diffey, H. R., UKAEA Rep. TRG 1256(W) (1966).

3

Radiation Hazards and Environmental Consequences of Reactor Accidents

J. R. Beattie

Radiation Hazards and Health
Physics Control Levels

The need for radiation protection was realized very soon after the discovery of X rays and radioactivity in 1895 and 1896, respectively, but the definition of practical measures required was not formalized until 1928, when the International Commission on Radiological Protection (ICRP) was established. This agency has regularly reviewed the problems of control of radiation hazards and issued its recommendations from time to time since then. These recommendations therefore embody today more than forty years of experience. Their observance has resulted in an almost total absence of radiation injuries among the many thousands who have worked in the burgeoning nuclear industry since the 1940s, and in the remarkably good health record of this industry in spite of the very large quantities of radioactive materials it has used and created. The current recommendations of the ICRP on maximum permissible doses for adults exposed to radiation in the course of their work are as follows [1]:

Whole body, gonads, red bone marrow	5 rems/yr
Skin, bone, thyroid	30 rems/yr
Hands, forearms, feet, ankles	75 rems/yr
Other single organs	15 rems/yr

The physical unit of absorbed radiation dose is the rad, defined as 100 ergs imparted to 1 g of tissue by ionizing particles generated by the radiation. The corresponding unit intended as a measure of potential biological damage is the rem, and for gamma or beta radiation 1 rad = 1 rem. For alpha radiation the dose in rems is obtained by multiplying the dose in rads by a factor of 10, which makes allowance for the potentially more damaging effects due to the greater linear density of ionization caused in tissue by alpha particles. More complex considerations apply in neutron dose estimation but in the context of public hazards from reactors neutron irradiation is not a significant risk, and need not be considered here.

The ICRP enjoins "that all doses be kept as low as is readily achievable, economic and social considerations being taken into account [1]." The dose limits considered appropriate, against which to control possible exposure of individual members of the public, that may arise from normal operation of nuclear installations (e.g., from low-level gamma radiation from the reactor, or indirectly from the controlled discharge of low-level gaseous and liquid effluents) are one tenth of the maximum permissible doses for radiation workers, with one exception mentioned below. The dose limits of most concern in practice are:

Whole body, gonads, red bone marrow	0.5 rem/yr
Skin, bone, thyroid	3.0 rems/yr
(but thyroid dose for children up to 16 yr of age, 1.5 rems/yr)	
Other single organs	1.5 rems/yr

National regulations are usually based on the ICRP's recommendations, but in the USA, for example, still lower limits governing the exposure of members of the general population are incorporated in national regulations. In the UK, lower limits are applied in practice, less formally, through the activities of the regulating ministries and their agencies. In practice and in order to conform to the ICRP's enjoinment, operators of nuclear installations take pains to ensure that any dose received by members of the public as a result of their operations does not exceed even a small fraction of the above dose limits. These are intended to minimize the risk of somatic effects of radiation affecting individual members of the general public, principally and in practice those few who live near the perimeter fence of nuclear sites. Somatic effects evidence themselves only at doses much higher than these limiting values. At very much higher doses they include damage to the blood-forming organs and to the intestines, for example, effects observable within hours or days of a high dose. Delayed effects include leukemia and other cancers, after a latent period measured in years. However, if radiation exposure of the individual is not allowed to exceed the

3. Radiation Hazards and Environmental Consequences

ICRP dose limits quoted above, it is in the highest degree unlikely that any such effects will be observed in addition to the natural incidence of disease in the absence of all radiation (except the natural background). In the UK, the system of government authorizations required before radioactivity can be discharged to the environment and the associated programs of environmental surveys and inspections have ensured that exposure of individual members of the public have been held at minimum values well below these ICRP limits and consistent with the demonstrable need for the discharges and application of the best practicable means for reducing them. In the USA, the principal of the ICRP that "all doses be kept as low as is readily achievable" has been translated into national regulations stipulating maximum dose rates for individual members of the public near nuclear sites which are a factor of ten or more lower than those listed above, generally on the grounds that these lower levels are readily achievable for current types of American water reactors. At the time of writing, the latest development is the publication by the US Environmental Protection Agency of proposed standards concerning radiation protection for nuclear power stations which would limit individual public exposures "near nuclear fuel cycle operations" to 25 mrems/yr whole body, 75 mrems/yr thyroid, and 25 mrems/yr to any other organ [3].

Irradiation of the gonads, which is one consequence of whole body exposure, implies a genetic risk to the population, particularly if large numbers of people are so exposed even at low doses consistent with the ICRP or other dose limits designed principally to control somatic effects. For this reason it is usual to state some limitation on whole body/gonad irradiation of the population on a national basis. ICRP [1, paragraph 86] recommends that this dose should "be kept to the minimum amount consistent with necessity and should certainly not exceed 5 rems" per generation (i.e., per 30 yr, since this is taken to be the mean age of child-bearing). This is intended to apply to exposure "from all sources additional to the dose from natural background and from medical procedures." National regulations usually embody limits within this ICRP criterion. The most recent proposal bearing on this in the USA is [3], which in addition to the dose limits mentioned above also puts forward limits on krypton-85 emission from the nuclear fuel cycle which have impact on the long-term whole-body dose to the population. National regulations introduced in the UK in 1960 are designed "to ensure, irrespective of cost, that the whole population of the country shall not receive an average dose of more than 1 rad per person in 30 years" from radioactive waste disposal. At the present time the average dose so received by the population of the UK is less than 1% of this limit.

All of the above-mentioned measures (the system of dose limits and the manner in which they are conformed to) reflect the fact that the original concept, that there were levels of dose or dose rates below which no injurious effect occurred, has given way over the last decade or so to the assumption that such effects may in some if not all cases be proportional to dose. There is evidence that the numbers of cases of various types of cancer induced by beta/gamma radiation is proportional to dose if this is greater than a few tens of rads. However, some effects, including for example radiation-induced cataract, show a threshold dose effect.

Dose limits should ideally be based on a balance between the benefit and risk of the work being undertaken, but such an ideal approach has not proved practicable so far. However, [2] has introduced proposed limits on waste discharges from nuclear plants and related dose limits for members of the US population, in the light of cost/benefit considerations. The difficulty is greatest when decisions have to be made on those dose levels which may be regarded as "safe" or "tolerable" in the event of an accidental release of radioactivity, for example, fission products from a reactor accident. Benefit could be quantified in terms of the electrical power produced, but risks from an accidental release are difficult to formulate in terms which would permit useful comparisons to be made.

The nature and degree of risk to life from nuclear radiations may be outlined as follows. The doses discussed in this paragraph and subsequently would not be permitted to occur during normal reactor operations. They could occur locally if there were a major accident to a reactor, although such an accident would be considered to have a very low probability of occurrence. It has long been known that 800 rads of whole-body radiation delivered over a short period would be lethal to all persons so exposed. Four hundred rads, the so-called LD_{50} dose, would be lethal to half the persons exposed. One hundred rads or less would have very little observable effect, at least in the short term. Whether long-term effects (principally cancer of one kind or another) would affect an exposed individual presumably depends to some extent on his age and state of health, but otherwise the evidence suggests that the induction of cancer by radiation is principally a matter of probability which increases with the dose delivered within certain roughly defined limits. There is a small definable probability that an individual who survives short-term injuries, if any, from a radiation dose may after a delay of years develop some form of cancer. The more sinister effects of radiation,

3. Radiation Hazards and Environmental Consequences

carcinogenesis in the individual and genetic risks following the exposure of populations, have recently been quantitatively assessed from observation of the survivors of the atomic bombing of Hiroshima and Nagasaki and of persons exposed to high radiation doses in the course of medical treatment. A report on this topic has recently been published by a United Nations committee [3], and its main conclusions have been summarized by Marley [4]. It was estimated that following short-term ("acute") exposure to whole-body radiation, the number of cancers of all kinds likely to arise thereafter is 250 per million man-rads, including some 50 cases of leukemia; this refers to an average population of average age distribution. Use of manrads as a measure of dose implies, for example, that 250 cases would result whether ten thousand people received 100 rads each or forty thousand people received 25 rads each. Following a single exposure of a group to radiation the number of cases of leukemia reaches a maximum six to seven years later, though a few cases may arise even after twenty years. For most other forms of radiation-induced cancer the mean latent period of induction is longer, some ten to twenty years or more. However, in the aftermath of a fission product release from a reactor accident, whole-body radiation is a relatively less important hazard than the selective irradiation of single organs by the inhalation or ingestion of certain isotopes. These organs include the thyroid gland which could absorb radioactive iodine either from inhalation of the contaminated air downwind of the release point or from ingestion through drinking contaminated cows' milk. Irradiation of the lung is also a possibility if insoluble long-lived fission products such as ruthenium were to be released and then inhaled. For such single-organ irradiation the number of cancer cases is of the order of 10 per million man-rads. Quite similar conclusions about the somatic effects and risks of radiation were reached by a committee of foremost US experts in this field and published in the BEIR report from the National Academy of Sciences [5]. These risks could in some cases be reduced by medical treatment. Leukemia, however, is almost invariably fatal. Lung cancer is unlikely to be curable unless detected early. The chance of curing thyroid cancer is thought to be as high as or even higher than 90% for young persons, but thyroid cancer is a more lethal disease in elderly people. It is important in analyzing risks to distinguish estimates of numbers of cancer cases from estimates of fatalities, which will usually be a lower figure.

On genetic effects of radiation, the United Nations committee [3] concluded that the natural incidence of genetically induced severely handicapping diseases was about thirty thousand per million live human births. The report considered the possible effect of exposing a large population to a low dose of radiation (additional to natural background) and estimates that the natural incidence of these diseases might be increased by 1% if the exposure of the large population averaged 1 rad/person. Genetic effects arise from irradiation of the gonads and therefore mainly, though not exclusively, from whole-body radiation. Since whole-body irradiation is a relatively minor aspect of the radiation dosage which might be expected to follow a hypothetical release of radioactivity from a reactor accident, genetic effects are unlikely to form part of the recognizable consequences to human health arising from a reactor accident release.

The UK Medical Research Council has from time to time published recommendations concerning levels of radiation and radioactivity in the environment and population in the aftermath of an accident to a reactor or other nuclear facility. The term "emergency reference level" has been coined in an attempt to describe more accurately the purpose of recommended levels of this type. According to Dunster [6], emergency reference levels (ERLs) are intended as guides for those who would be responsible for initiating countermeasures if an accidental release should occur. If the doses incurred are likely to fall below the appropriate ERL, it is thought that countermeasures should be considered and put into effect if they will significantly reduce the dose without adding to the risk in some other way. It should be borne in mind that the ERL carries with it only a very low risk that people affected might be hurt: in most cases a risk of the order of 1 per 1000 that they might die as a result in the next 10-20 years. This is, then, a risk of about 1/10,000 per year, some five times lower than the average risk of accidental death. ERLs are given in Table I for iodine-131 and cesium-137, the isotopes of greatest importance in relation to fission product release from present-day thermal neutron reactors, for ruthenium-106 which would probably be the principal cause of a dose to the lung if this were significant in these circumstances, and also for strontium-90 which would be the principal cause of a dose to bone and bone marrow [7].

The ERL of external dose from gamma radiation to be taken into account when considering possible exposure of the public over a short period following a reactor accident release would be 10 rads average in body tissues, which corresponds approximately to 15 R measured in free air.

The protective action guides recommended by the US Federal Radiation Council for use by federal agencies in radiation emergencies may also be considered at this point. Some values recommended by the US Federal Radiation Council are given below

3. Radiation Hazards and Environmental Consequences

TABLE I

Emergency Reference Levels for Iodine-131, Cesium-137, Ruthenium-106, and Strontium-90

Isotope	Critical organ	ERL of dose (rems)	ERL of cloud-dosage (Ci-sec/m^3)	ERL in milk (µCi/liter)
Iodine-131[a]	Thyroid	30	0.020	0.25
Cesium-137[b]	Whole body	10	1.5	5.5
Ruthenium-106[c]	Lung	30	0.014	—
Strontium-90	Bone marrow	10	0.05	0.15

[a] ERL of cloud-dosage makes allowance for iodine and tellurium in equilibrium proportions.
[b] External gamma radiation from cesium-137 deposition is more limiting than the ERL in milk; see text.
[c] Ruthenium is released as insoluble ruthenium oxide, and does not appear in cows' milk.

TABLE II

USFRC Protective Action Guides

Isotope	Critical organ	Dose (rads)	Concentration in milk (µCi/liter)
Iodine-131	Thyroid	30	0.18-0.21
Cesium-137	Whole body	10	2.4
Strontium-90	Bone marrow	15	0.155
	Mineral bone	45	

in tabular form (Table II), permitting comparison with the UK values of ERL given above. These FRC data for iodine-131 are taken from [8], and for cesium-137 and strontium-90 from [9]. It will be seen that the Federal Radiation Council's data on protective action guides are generally similar to the British Medical Research Council's data, although some differences in detail are evident.

Reactor Accidents and Fission Product Release to the Atmosphere

There is general agreement that nuclear power reactors must be designed, constructed, and operated in such a manner that uncontrolled releases of radioactivity should not occur.

In spite of every care accidents can still happen; there will always be a finite but small probability that an uncontrolled release of radioactivity to the atmosphere could occur, which would adversely affect the health and safety of some members of the public. For the present purpose, discussing thermal neutron reactors, we need not make distinctions, invidious one way or the other, between gas-cooled and water-cooled reactors. They all contain fission products in amounts primarily proportional to their thermal power (in the case of long-lived isotopes such as cesium-137, proportional also to the length of time the fuel has been irradiated in the reactor). The fission products likely to be released from fuel which is damaged in an accident are those which are gaseous or readily volatilized, because the type of damage most likely to occur is overheating. Some of the volatile and, therefore, also recondensable fission products may never reach the outside atmosphere because they can be deposited within the reactor and its enclosing buildings, and some may be removed by filter absorbers installed for this purpose along gas outlet routes. For the present let us ignore these effects which will vary from system to system and consider the worst state of affairs, for simplicity if for no other reason, in which the predominantly gaseous and volatile fission products are released from some limited and arbitrary quantity of fuel directly to the atmosphere. What is likely to be the composition of the radioactive fission product mixture released in these circumstances? Some evidence is available from accidents in the now distant past to reactors of types now obsolete, and shutdown or demolished. As for the power reactors in use today, there is evidence from laboratory experiments in which fission products were released from fuel specimens which were irradiated in a reactor and then removed and deliberately overheated in a laboratory rig. From this evidence, most of which has been reviewed [10], one foresees that isotopes of xenon, krypton, iodine, cesium, and possibly also ruthenium (if oxygen is present and the temperature is high enough) are likely to predominate in the release from fuel; they will probably be present in that release in relative proportions very similar to those in which they were present in fuel just before the accident occurred. This rather general conclusion is on the whole well borne out by the results of the theoretical "Reactor Safety Study" recently published in draft form by Rasmussen and his co-workers [11].*
This is an elaborate assessment of accident risks in two typical US commercial nuclear power plants, one a pressurized

*The final version issued in December 1975 does not differ significantly in this respect.

3. Radiation Hazards and Environmental Consequences

TABLE III

Components of a Nominal Release of Gaseous and Volatile Fission Products Containing 10^3 Ci of Iodine-131

Component	Activity (Ci)
Short-lived gases and volatiles	1×10^5
Short-lived gases	2×10^4
Iodine-131	1×10^3
Cesium-137 (after 1-1/2-yr irradiation)	50
Ruthenium-106 (if all is released)	5×10^2
Strontium-90 (if 5% is released)	4

water reactor, the other a boiling water reactor. There are found to be nine possible release categories for PWRs and six for BWRs—most of the detailed analysis relevant to this discussion is to be found in Appendixes V, VI, and VII of Ref.[11]. Table VI-2 in Appendix VI sets out the derived release fractions from fuel inventory to the free atmosphere, for the fifteen accident categories according to chemically similar groups of fission product isotopes. For those accidents of a more catastrophic nature (and also of very low probability), Table VI-2 shows quantitative releases of noble gases, iodine, cesium, and tellurium. These large releases, in some cases, involve quantitative release of ruthenium, though in most cases there is forecast only a moderate release of ruthenium. A moderate release of strontium is also predicted for these large releases. One may bear all this in mind in considering the simplified pictures of the events which follow.

It is common practice, particularly in regulatory circles, to consider the effects of a nominal or arbitrary release of gaseous and volatile fission products which includes in its composition 1000 Ci of iodine-131. The inventory of iodine-131 in a present-day thermal neutron reactor of 600 MWe and 1500 MWt is approximately 5×10^7 Ci, so that no particular significance should be attached to the figure of 10^3 Ci. Nevertheless, it has the merit that it is a small fraction of the total which one would hope at least not to exceed in a foreseeable, if unlikely, accident; but this can by no means be guaranteed. Let us examine the probable consequences to public health of this release, which would in broad outline be made up of the following components (Table III). Strontium, usually present as strontium oxide, is not usually volatile, but in view of recent findings [11], a moderate release of this element is assumed. Many other fission products, such as those in the rare earth group, are highly refractory and involatile, and are unlikely to form more than a very small part of the release. Reference [11] includes plutonium in this very low release group.

The release proportions which national regulations in various countries require applicants for reactor licenses to consider are seldom specified in so much detail, but are generally similar to the above. The first analysis of the possible consequences to public health from fission product releases to be published appeared in a paper presented by Marley and Fry to the Geneva conference of 1955 [12]. In this the authors considered the release of a full spectrum of both volatile and nonvolatile fission products, and focused attention on whole-body and lung irradiation arising from mainly insoluble airborne radioactive material and on contamination of inhabited areas and the environment with deposited radioactivity. In 1957 a study team of USAEC staff considered the problem in further detail and drew attention to the likely features of a release consisting predominantly of those fission products which are gaseous or volatile at moderate temperatures [13]. The release of iodine-131 which occurred in October 1957 as a result of the fire in the air-cooled reactor at Windscale drew attention to the possible importance of iodine isotopes as a source of hazard in reactor accidents [14], and stimulated the publication of a number of reports in which the environmental hazards of accidental releases of volatile fission products were assessed with the object of providing technical guidance to those formulating legislation and administrative procedures for the control of any such future emergencies. Numerous reports of this type have appeared in the United Kingdom, United States, Germany, and elsewhere [15,16,17]. Subsequently, some of the assumptions typically made in these reports found their way in simplified form into official regulations and guides. For instance, a paper outlining experience of reactor siting in the UK, published by the nuclear inspectorate staff [18], based its considerations on the consequences of a release of 10^3 Ci of iodine-131 in Pasquill Class F weather conditions. In the US, guidance to applicants for construction permits or operating licenses for light water reactors included a statement of the regulatory position to the effect that 25% of the equilibrium iodine inventory should be assumed to be available for leakage from the primary reactor containment, together with 100% of the noble gas inventory [19]. Although basic release assumptions differ in detail, there is, nevertheless, a basic similarity of approach to the analysis of hazards from potential accidental fission product releases as carried out in the various countries concerned in developing nuclear power.

3. Radiation Hazards and Environmental Consequences

A full discussion of how the components of release would be carried downwind and dispersed first by building-induced turbulence and later by small and large-scale natural turbulence in the atmosphere is clearly beyond the scope of this monograph. The phenomena are complex and many meteorologists such as Sutton, Pasquill, and Gifford have contributed to research on them [20]. In the weather conditions which occur most frequently in the UK and many other countries, the concentration on the axis of the plume will vary with distance x downwind approximately as x^{-2}. These are the moderately turbulent conditions associated with zero or small vertical gradient of atmospheric potential temperature (temperature adjusted for the adiabatic expansion of air with increasing height above ground) which occur some 60% of the time in the UK. An average figure for the US is about 40%, as one may deduce from data given in Table VI-4, Appendix VI of Ref. [11]. For about 20% of the time and mainly during the night, moderate to strong inversions of potential temperature gradient occur, and in these conditions of low turbulence and low wind speed, concentrations are a factor of 10 or more higher than in the more turbulent conditions just mentioned. This is true for the UK, most of the US, and many other countries. In this type of weather the concentrations of a gas or nondepositing aerosol released at or near ground level varies with distance roughly in proportion to $x^{-1.5}$; but for depositing aerosols, such as elemental iodine and fine particles with aerodynamic diameters in the range 0.1-10 μm, the plume concentration in air would vary with distance approximately as x^{-2}. The difficulties in making precise predictions of aerosol concentrations and consequent ground deposition are great, but a general picture is readily obtained. It is as well to note, also, that the concentration from a given release at a given distance downwind may vary considerably depending on the weather; and that the probability of occurrence of a given weather condition varies somewhat from year to year.

If we imagine that the release of fission products tabulated above has occurred from a point near ground level and in the most probable weather type in the UK, we can calculate and describe the consequences to the public and their environment downwind of the reactor as follows.

(i) Whole-body radiation doses from external gamma rays from the cloud of gaseous and volatile fission products would be a few rads on parts of the reactor site, but off the site this radiation component would be very small, indeed less than 1 rad and quite insignificant as a hazard. Although such external radiation from the cloud would be negligible, internal radiation received by some members of the public who, situated in the path of the cloud, inhaled the contaminated air would be a significant hazard. However, inhalation of cesium-137, which leads to a whole-body dose, would give only 9 mrems at 1.6 km (1 mile) downwind—a trivial dose at this distance.

More important is the inhalation of iodine; iodine rapidly passes through the lung into the bloodstream from which about 30% is absorbed and concentrated in the thyroid gland over a period of several hours. The dose delivered in this way to the thyroid gland of a child of about one year of age at 1.6 km (1 mile) downwind from the reactor would be about 20 rems, i.e., just less than the ERL in the UK and the protective action guide in the US (both 30 rems). Thyroid doses at other distances can be estimated from the approximate inverse square law already mentioned. The thyroid dose to an adult at the same distance would be a third of the dose to the young child. The principal form of risk to a member of the population who receives a dose to the thyroid is that he may subsequently develop thyroid cancer. However, in the case of a 1000-Ci release, as considered here, it is unlikely that a single case of thyroid cancer would arise even on the more populated sites most recently used in the UK. These have a distribution of population about 5-10 times higher than the average of the first sixty sites chosen in the US. Indeed, even for these populated sites the probability that one case of thyroid cancer would arise following this release would be of the order of 1 in 10. Remedial action is possible which can reduce the thyroid dose by a factor of 10 or more and can reasonably be applied to those people most at risk. If a tablet containing 100 mg of stable potassium iodide or iodate (200 mg in the case of an adult) is taken by mouth as soon as possible after, and at most within 2 hr of, the inhalation of radioactive iodine, this rapidly satisfies the thyroid gland's requirement for iodine and ensures that most of the radioactive iodine which might otherwise have gone to the thyroid is rejected in urine. This remedial action is included in UK plans for such an emergency.

If some ruthenium-106 is released from the reactor, this will be in the form of ruthenium oxide which may be inhaled if the particles are sufficiently small (i.e., have an aerodynamic diameter less than about 10 μm). Ruthenium oxide is "insoluble" in lung fluids and it has been shown by several investigators that its biological half-life in the lung is 230 days. If inhaled, it and its short-lived rhodium daughter deliver a dose primarily to the lung, which would be about 14 rems to the young child, if one assumes that ruthenium is as readily released from the reactor as is iodine. If a lower release fraction is assumed for ruthenium, e.g., 5%-10%, then

3. Radiation Hazards and Environmental Consequences

the dose to the lung would be about 1 rem, and this is probably a more realistic estimate. This dose to the lung is still quite small for the size of accident considered.

The release of 4 Ci of strontium-90 in the conditions mentioned would lead to a dose to bone marrow arising by inhalation. For a child 1 mile downwind this would be only 10 mrems.

(ii) Iodine would be deposited on pastures and other land and buildings downwind. If nothing were done, children (and adults) could continue to drink milk now contaminated with iodine-131 at or above the ERL; this would make it desirable to ban the consumption of milk arising from the affected pasture land over an area 10-15 km downwind and several kilometers wide. The ban would last for several weeks and the cost of compensating milk suppliers in typical country surrounding UK reactors would be of the order of £10,000. Ingestion of iodine and other activity through eating exposed green foods and through inadvertent ingestion during normal living would probably be a problem, in the example cited, only as far as 500 m downwind.

Close to the reactor, within about 150 m, people would be exposed to significant levels of external gamma radiation from iodine isotopes deposited on the ground. This radiation would persist for 3-4 weeks, during which time any person continuously out-of-doors in that area would accumulate about 10 rads, which in the UK is one ERL of exposure. It is unlikely that any members of the public would be affected at such a short distance from the reactor, but if the release were bigger and a larger area were affected, remedial action would consist of temporary evacuation of local residents. After 4 weeks or so had passed and gamma radiation from iodine deposition had decayed considerably, long-term low levels of gamma radiation from barium-137, the short-lived daughter of deposited cesium-137, would become more apparent. For the tabulated release, the level of deposited cesium-137 at 1 km downwind would be about 10^{-5} Ci/m^2 and the gamma dose rate therefrom would initially be 0.5 rad/yr. Cesium-137 has a radioactive half-life of 30 yr, and it is known that cesium becomes entrapped in the crystal lattices of clay minerals [21]. This process appears to occur in 1-2 yr, during which time the cesium moves downwards a few centimeters in the soil profile; as a result the initial gamma dose rate declines by about half in the first year after deposition, and thereafter declines more slowly with an apparent half-life of about 20 yr [22]. Because 0.5 rad/yr is the ICRP dose limit for whole-body radiation exposure of individual members of the public, it is difficult to dismiss such long-term exposure as unimportant. It might well interfere with the occupation and use of land and buildings unless these could be decontaminated in some way.

The hazards from ground deposition are long term in nature and their full effect depends on integrated irradiation or uptake over a period of days, weeks, or even longer following the accident. There is, therefore, more time for effective countermeasures to be put into practice than in the case of the hazards which proceed directly from the cloud of radioactivity to the person.

The above account describes the consequences of a release containing a nominal 1000 Ci of iodine-131. To cope with the increasing number of environmental hazard analyses required by licensing and other authorities and especially for the assessment of potential radiological hazards from nuclear accidents, many nuclear centers and reactor manufacturers have developed computer programs. The compilation, mainly of US codes, made by Winton includes 132 computer codes for nuclear accident analysis [23]. In the UK, an outstanding example of such a code is the WEERIE program compiled by Clarke [24].

Since a reactor operating at 1500 MWt contains 5×10^7 Ci of iodine-131 in all, it would be entirely reasonable for the concerned reader to inquire about the likely effects of a catastrophe in which say 10^7 Ci of iodine-131 among other gaseous and volatile fission products were released. An approximate answer can be obtained at once from the inverse square law of distance (i.e., concentration proportional to x^{-2}) previously mentioned. It follows from this that a release which is a factor of 10^4 times greater than our example, namely 10^7 Ci of iodine-131, would produce the consequences just described at distances approximately 100 times greater. For example, the dose to the child's thyroid by inhalation so reckoned would be 20 rems at approximately 100 miles, and gamma radiation from deposited iodine would apparently be of concern at 15 km while gamma radiation from deposited cesium-137 could be significant up to about 100 miles. Two factors should prevent the full realization of these consequences. The most important of these comprises the quite outstanding attention given to component quality and reliability, operator training, and intrinsic design features—in short to safety—in all reactor work. A secondary safety feature of such large releases is the heat generated by the release by virtue of its own radioactivity. In all, this hypothetical release would contain about 10^9 Ci of assorted activity and this radioactivity would release heat at the rate of more than 5 MW within the plume. That the plume will rise seems to be an inevitable consequence, and if this is so then most of the expected effects at ground level will not be realized. The theory of

3. Radiation Hazards and Environmental Consequences

this mode of plume rise has been developed by Gifford [25]. Great caution has been shown in the nuclear industry, which has generally not so far accepted the reality of this type of plume rise [11].

One other query of considerable importance is whether it is conceivable for the fuel itself and the refractory and normally nonvolatile fission products such as zirconium, niobium, cerium, and so forth, to be vaporized and released. These fission products comprise some 75% of fission product activity and many of them have radioactive half-lives of several months. The fuel itself may contain plutonium which would be released as insoluble plutonium dioxide if such a vaporization could occur. The consequences to public health of such a vapor release may be put in focus by imagining the vaporization and release of all the fuel and all the fission products in fuel elements containing the 1000 Ci of iodine-131 used in our previous example [26]. The chief new components added to the release would be about 2×10^4 Ci of relatively long-lived and insoluble fission products, and up to about 35 g of plutonium and other transuranic elements. Among the principal health consequences would be a dose to the lung of a child at 1.6 km (1 mile) amounting to about 100 rem, part due to insoluble fission products and part due to plutonium and other alpha-emitting transuranics. Gamma radiation from ground deposition would be hazardous at the same distance—about 100 rads/yr from iodine and rare earth isotopes principally, which would decay in 2-3 yr leaving a residual radiation field of order 0.1 rad/yr due to deposited cesium-137. Plutonium deposition at this distance would constitute a long-term hazard—if one assumes a typical range of deposition velocities from plutonium aerosols $3-10 \times 10^{-3}$ m/sec, the level of ground contamination by plutonium would be 3-10 $\mu g/m^2$. If only plutonium-239 were present, the alpha emission from the ground would be $2-6 \times 10^{-7}$ Ci/m, which is 2-6 times greater than the highest level of alpha emission permitted in "inactive and low activity" areas in the laboratories and factories of the UK Atomic Energy Authority and would be unacceptable in an area used by the general public. Presumably the area would have to be evacuated or decontaminated. In relation to both the lung dose due to plutonium and to ground contamination by alpha emitters, it is worth noting at this point that if the fuel involved in the accident were high burn-up fuel, the specific activity of alpha emission would be increased several-fold over that of plutonium-239, because of the presence in the fuel of alpha-emitting transuranics of shorter half-life and higher specific activity than plutonium-239. The lung dose from plutonium and transuranics in reactor accidents depends almost entirely on their total alpha activity.

The lung dose from inhaled plutonium-239 may be calculated using ICRP data [27], and a recent publication of the British Medical Research Council gives specific consideration to the lung dose from plutonium inhaled in accident conditions [7]. As for other radioactive isotopes, the ERL of dose to lung is taken to be 30 rems in [7], and it is shown that this dose to a child's lungs would result from exposure to a cloud-dosage of 1.5×10^{-4} Ci-sec/m^3 of plutonium-239. In the example discussed at some length in the previous paragraph, the release of 35 g of plutonium corresponds to 2 Ci of plutonium-239, and the lung dose to a child at 1 mile would be about 5 rems from this isotope. If the fuel were of high burn-up, other alpha-emitting transuranics along with the plutonium-239 could raise this part of the lung dose to 20-30 rems. The remainder of the lung dose would be delivered by moderately long-lived and insoluble fission products such as ruthenium-106 and cerium-144. One other aspect of the consequences of inhaling "insoluble" plutonium should be mentioned, namely, the dosage delivered in the long term to liver and bone by particulate plutonium which is transferred in the course of time from the lungs to these organs. It has been estimated that the total "risk" to a person inhaling plutonium-239 might be increased by a factor of about 3 in this way, and some allowance for this might, on further consideration, have to be made in the ERL of cloud-dosage [28].

All of these considerations about the effects of a release of vaporized fuel merit serious attention. But one may ask: "What is the likelihood of such a release of vaporized fuel occurring?" If the answer is "none," our conception of the effects of a release as limited to the effects of noble gases and volatiles such as iodine would not be affected. Fortunately, the response of thermal reactors to the range of reactivity additions that are physically plausible would be controlled by rapidly acting temperature coefficients of reactivity in a way which would preclude fuel vaporization and release of this kind—in short, the fuel would melt and rearrange itself so as to shut down the nuclear reactor before vaporization could intervene. In fast reactors, also, reactivity additions must be considered and, with the reactivity coefficient available, fuel vaporization is theoretically possible given very rapid reactivity addition. Fast reactor design is purposely arranged to exclude this possibility and the reactor containment is of a kind designed to prevent any release.

3. Radiation Hazards and Environmental Consequences

References

1. ICRP. Publication 9, "Recommendations of the International Commission on Radiological Protection, Adopted September 17, 1965," Pergamon Press, New York, 1966.
2. Environmental Protection Agency. Radiation Protection for Nuclear Power Operations—Proposed Standards, Federal Register Vol. 40, No. 104 (May 29, 1975).
3. United Nations. Ionizing Radiation Levels and Effects, A Report of the United Nations Scientific Committee on the Effect of Atomic Radiation, New York (1972).
4. Marley, W. G. Atomic energy and the environment, introductory paper presented to the NEA/IAEA Symposium on the Management of Radioactive Wastes from Fuel Reprocessing, Paris, November 27, 1972.
5. National Academy of Sciences, National Research Council. The Effects on Populations of Exposure to Low Levels of Ionizing Radiation, Washington, D. C. (November 1972).
6. Dunster, H. J. The Application and Interpretation of ICRP Recommendations in the United Kingdom Atomic Energy Authority, UKAEA Rep. AHSB(RP)R78, HMSO (1968).
7. Medical Research Council. Criteria for Controlling Radiation Doses to the Public after Accidental Escape of Radioactive Material, HMSO, London (1975).
8. US Federal Radiation Council. Radiation Protection Guidance for Federal Agencies, Federal Register Vol. 29, No. 165, p. 12057 (August 22, 1964); see also Federal Radiation Council Staff Rep. No. 5, Background Material for the Development of Radiation Protection Standards (July 1964).
9. US Federal Radiation Council. Radiation Protection Guidance for Federal Agencies, Federal Register (May 22, 1965); see also Federal Radiation Council Staff Rep. No. 7, Background Material for the Development of Radiation Protection Standards: Protective Action Guides for Strontium-89, Strontium-90 and Cesium-137 (May 1965).
10. Beattie, J. R., and Bryant, Pamela M. Assessment of Environmental Hazards from Reactor Fission Product Releases, UKAEA Rep. AHSB(S)R135, HMSO (1970).
11. Rasmussen, N. C., et al. Reactor Safety Study—An Assessment of Accident Risks in U.S. Commercial Nuclear Power Plants, USAEC Rep. WASH-1400 (August 1974, December 1975).
12. Marley, W. G., and Fry, T. M. Radiological hazards from an escape of fission products and the implications in power reactor location, Proceedings of the First International Conference on the Peaceful Uses of Atomic Energy, Geneva, 1955, Paper A/CONF.8/P/394.

13. USAEC. Theoretical Possibilities and Consequences of Major Accidents in Large Nuclear Power Plants, USAEC Rep. WASH-740 (1957).
14. Report of the Committee of Enquiry. Accident at Windscale No. 1 Pile on 10th October, 1957, Cmnd 302, HMSO, London (1957).
15. Beattie, J. R. An Assessment of Environmental Hazard from Fission Product Releases, UKAEA Rep. AHSB(S)R64 (1961).
16. Dinnunno, J. J., et al. Calculation of Distance Factors for Power and Test Reactor Sites, USAEC Rep. TID 14844 (1962).
17. Vogt, K. J. Dispersion of Airborne Radioactivity Released from Nuclear Installations, paper IAEA-SM-181/39, Vienna (November 1973).
18. Charlesworth, F. R., and Gronow, W. S. A summary of experience in the practical application of siting policy in the United Kingdom, IAEA Symposium, Vienna, 1967, paper IAEA-SM-89/41.
19. Division of Reactor Standards, USAEC. Assumptions Used for Evaluating the Potential Radiological Consequences of a Loss of Coolant Accident for Boiling Water/Pressurized Water Reactors, Safety Guides 3 and 4 (1970).
20. Slade, D. H. (Ed.) Meteorology and Atomic Energy. USAEC Division of Technical Information (July 1968).
21. Scott-Russell, R. (Ed.) "Radioactivity and Human Diet," Chapter 14, Pergamon Press, New York, 1966.
22. Gale, H. J., et al. The Weathering of Caesium 137 in Soil UKAEA Rep. AERE-R4241 (1963).
23. Winton, M. L. A compilation of computer codes for nuclear accident analysis, *Nucl. Saf.* 10, 131 (1969).
24. Clarke, R. H. A Users' Guide to the Weerie Program, CEGB Rep. RD/B/N2407 (April 1973).
25. Gifford, F. A. The rise of strongly radioactive plumes, *J. Appl. Meteorol.* 6, 644-649 (1967).
26. Beattie, J. R. Some Considerations of the Effects of the Accidental Release of Fission Products from Reactors, Symposium on Environmental Surveillance around Nuclear Installations, Warsaw, November 1973, paper IAEA/SM-180/6.
27. International Commission on Radiological Protection, Rep. of Committee II. "Permissible Dose from Internal Radiation," Pergamon Press, New York, 1959.
28. Medical Research Council. The Toxicity of Plutonium, HMSO, London (1975).

4

The Calculated Risk—A Safety Criterion

G. D. Bell

Risk is inherent to the human condition. There are
natural risks to which all are prone. These include the possibility of illnesses, including cancers, resulting in death.
Other natural risks may arise from the environment; floods,
high winds, lightning, and earthquakes regularly take their
toll in human society. Civilization brings about a redistribution in hazards. Advances in medicine enable more people
to reach old age so that the average life span, which did not
change much between Roman times and the end of the eighteenth
century, has since been almost doubled. Development of transport has introduced new risks, as has the pollution caused by
industrial activity. The use of atomic energy promises a
large additional contribution to the world's energy supply
at a time when other resources are becoming strained, but as
with all industrial processes, there are risks associated
with its development. In contrast to most new techniques,
the hazards arising from the use of nuclear power were recognized without the experience of disasters to demonstrate the
need for safeguards. The nature of the hazard, exposure to
radioactive materials, is a new one, and both because of its
insidious nature (the inability of the senses to detect or
appreciate the presence of radiation) and the world's first
experience of nuclear power in war, there has been a continuing public unease about operations involving radioactive
materials. That the risk, so far as the public is concerned,
has thus far remained potential rather than real has not allayed the anxiety of some people. It is thus clearly necessary that those concerned with the safety of nuclear plants

TABLE I

Some Values of the Probability of Death per Person per Year

Risk of death from all accidental causes	8×10^{-4}
Traffic accidents	2.5×10^{-4}
Falls in the home	
All ages	1.7×10^{-4}
0–45 yr	0.4×10^{-4}
45–65 yr	3.6×10^{-4}
Above 75 yr	2.6×10^{-3}
Leukemia from natural causes	5×10^{-5}
Death from thyroid cancer arising naturally	$\sim 10^{-6}$ (depending on age)
Struck by lightning	5×10^{-7}
Total probability of death within 1 yr	5×10^{-4} age 10
	1.1×10^{-3} age 20
	1.3×10^{-3} age 30
	2.6×10^{-3} age 40
	7×10^{-3} age 50

win acceptance for the standards of safety to be set and achieved, not only by the specialists within the industry but by the informed public as well.

    No human activity is without risk, and nuclear power is no exception to the general rule. The nature of the risk to the individual from accidental releases of radioactive material has been described in Chapter 3 and takes the form (for doses less than those sufficient for rapid lethality) of a statistical probability that cancers may be induced some years after exposure. Since no guarantee of absolute safety can be given, in this sphere any more than in any other, it is necessary to attempt to assess the magnitude of the risk being presented and to put it into context with other risks to life. Only after such an exercise can society make an informed and balanced judgment of the acceptability of nuclear power.

    The overall risk of death is a function of age, being high in the first year of life and again in the years of old age. From early manhood to early middle age the risk rate in technically advanced societies is in the range of 1–3 × 10^{-3}/yr and thereafter increases rapidly.

    The overall risk is compounded of many separate hazards, some of which with their corresponding risk rates (that is to say the probability of death per year) are given in Table I [1].

4. The Calculated Risk—A Safety Criterion

TABLE II

Summary of Estimates of Annual Whole-Body Dose Rates in the United States (1970)

Source	Average dose rate^a (mrems/yr)	Annual person rems (in millions)
Environmental		
Natural	102	20.91
Global fallout	4	0.82
Nuclear power	0.003	0.0007
Subtotal	106	21.73
Medical		
Diagnostic	72^b	14.8
Radiopharmaceuticals	1	0.2
Subtotal	73	15.0
Occupational	0.8	0.16
Miscellaneous	2	0.5
Total	182	37.4

^{a}The numbers shown are average values only. For given segments of the population, dose rates considerably greater than these may be experienced.
^{b}Based on the abdominal dose.

It is of interest that in children aged 1-14 accidents claim more lives than the five leading diseases combined, while in youths aged 15-24 accidents claim more lives than all other causes combined, motor vehicles representing 70% of the total accidental deaths. This represents a risk rate of 4×10^{-4}/yr. Indeed, accidents remain a dominant cause of death until middle age, when the incidence of disease rises rapidly. For example, the mean risk rate for heart disease alone in the age range 45-64 yr is about 4×10^{-3}/yr.

The operation of a power station imposes some additional risk to the surrounding population, and involves risk to workers supporting the plant (e.g., coal or uranium miners) and those who operate it. Daily effluent release from nuclear power stations is small and within ICRP recommendations for the nearest population. A summary of annual whole body dose rates in the United States (1970) has been given [3] and is reproduced here as Table II.

Major contributors to radiation dose are natural background and medical applications, the latter representing by far the greatest portion of manmade radiation. The present

TABLE III

Estimated Deaths per Thousand People, Each Working for 40 Years and Receiving an Occupational Exposure of 1 rem/yr

Leukemia	1
All other cancers	5
All other somatic effects, including risk of cardiovascular disease and renal disease	Small

contribution from nuclear power is trivial compared with natural background, and is expected to remain below 1 mrem/yr even with the expansion of nuclear power to 800,000 MW expected in the US by the year 2000 [4]. This represents less than 1% of natural background, which itself varies much more considerably from place to place. The cosmic ray component, for example, increases by a factor of three from sea level to 10,000 ft and also varies with latitude. In the US, the cosmic ray component varies from 38 mrem/yr in Florida to 75 mrem/yr in Wyoming. The terrestrial component shows a similarly wide variation according to locality. It would not appear to be logical to ignore such natural variation and yet to be unduly perturbed about the operational contribution from nuclear power.

The risk to operators has been estimated by Sowby [5] on the supposition that risk is proportional to dose even at low doses and dose rates. There is no evidence either to support or to refute this assumption. The results of this estimate of cases of cancer are given in Table III. It represents the lifetime risk arising from a nominal annual exposure of 1 rem throughout the body each year for 40 years.

The risk of death from accidents at work for an individual working for 40 yr in a given occupation are shown in Table IV. The first five occupations are among those generally considered to be hazardous and the lifetime risk for these exceeds 1%. The risk for all manufacturing industry lies at about 0.5% or rather less. The 0.3% risk for the USAEC is for risks not induced by radiation.

Finally, it should be pointed out that Table IV is based upon actual deaths, while the radiation risk remains an estimate of possible cancer induction. The most recent health statistics of the UKAEA [6] show that the cancer risk among its staff and pensioners is somewhat less than that expected for the general population, but this can hardly be taken to imply that occupational exposure effects an improvement in cancer risk, since it may be argued that the group is not typical and is subject to special medical attention. However,

4. The Calculated Risk—A Safety Criterion

TABLE IV

Deaths per Thousand People, Each Working for 40 Years in the Designated Occupation

Occupation	Deaths per thousand	
	UK	US
Trawler fishing	50	—
Aircraft crews (civilian)	70	—
Coal mining[a]	20	50
Pottery (pneumoconiosis)	20	—
Construction	50	—
All manufacturing	3	5
All industries	—	10
USAEC	—	3

[a] Including mining and quarrying.

with all the qualifications it is of some interest that the potential risk from occupational exposure remains unsubstantiated by experience.

The major area of concern then is associated with the possibility, though remote, of an accidental release substantially in excess of the permitted daily discharges. Two types of consequence may be distinguished arising from doses directly received from the cloud as it passes downwind. The first is the risk to an individual in the community. Individual risk rate clearly depends upon proximity to the station, other things being equal. If, therefore, the individual risk for someone living close in can be demonstrated to be acceptably low, the risk to others will be even less. The second consequence is what may be termed a community risk. Even though the individual risk may be shown to be sufficiently low, the community may be expected to be particularly concerned about events giving rise to large numbers of "casualties" (cases of cancer) within a large population. Most of these "casualties" would be estimated cases of cancer predicted to occur 10-20 yr after exposure. Such accidents are also likely to give rise to the need for local evacuation and for restrictions on the consumption of exposed food and milk produced from contaminated grassland. The immediate and unavoidable hazard, however, arises from inhalation of and exposure to radioactive material from the plume as it moves downwind.

Accidents to thermal reactors involve the melting or overheating of fuel, releasing the more volatile fission product isotopes and the inert gases. As described in Chapter 3, such a cloud gives rise to doses from inhalation primarily to the thyroid, with somewhat smaller doses to the lung and much smaller doses again to the whole body from external radiation. The ensuing risk rates of cancer developing in the population are then taken to be [7,8]:

Thyroid cancer $\qquad$ $10\text{-}30 \times 10^{-6}$ per rad according to age

All cancers from whole-body irradiation $\qquad$ 2×10^{-4} per rad

At short distances from the reactor, that is, up to about 2 miles, the value for the thyroid dose/whole-body dose is about 250, assuming a 100% release of all iodine isotopes and inert gases as present in the fuel to the outside atmosphere in average weather conditions. The ratio is higher for inversion weather conditions and increases with distance from the reactor.

Taking the lowest ratio and the cancer risk rates shown above, induced thyroid cancers would exceed all cancers induced by the whole-body dose by a factor of about ten. If the lethality of thyroid cancer were to be 10% (and this is debatable and may be lower), the number of deaths from thyroid cancers would be about equal to those from the whole-body dose. To avoid too much complexity in the discussion of risk then, we may take the incidence of thyroid cancers as a measure of the risk arising from releases of volatile fission products, bearing in mind that there may be circumstances in which the fractional release of isotopes differs from that which we have taken to be typical. The size of release will therefore be denoted by quoting its ^{131}I content, as this isotope makes the biggest contribution to the thyroid dose.

With this simplification, it is possible to estimate the risk to individuals in the event of an accidental release. For example, consider the release of 1000 Ci of ^{131}I at ground level and in average weather. The dose to the thyroid of a child, from inhalation, living at a distance of 1000 m downwind would be 10 rads, and assuming linearity of the dose/risk relationship at this low dose level, the risk of thyroid cancer developing at some future date would be 3×10^{-4}. It should be noted that this is the total risk, integrated over life. There is normally an induction period before cancers develop after irradiation and their appearance is then spread over a further period of time. To compare the risk rate following the accident with the other risks to life, it is necessary to divide the total risk by the number of years over which there is a continuing risk. If we take 10 yr as a conservative estimate for this, the annual risk rate for cancer development from the accident becomes 3×10^{-5}, and the death rate about 1/10 of this, i.e., about 3×10^{-6}/yr.

4. The Calculated Risk—A Safety Criterion

It will be noted that to arrive at this estimate it was necessary to specify the size of the release and the type of weather and wind direction at the time. In considering the risk from a nuclear reactor, there is an obvious difficulty here, in that a range of accidents is possible in a plant of such complexity. The total quantity of ^{131}I in a reactor of 1500 MWt power is about 4×10^7 Ci, and to isolate one release as "typical" would be a great oversimplification. Experience offers little assistance since, although the total accumulated operating experience of power reactors amounts to several hundred reactor years, there has been no accident involving significant release of activity to the environment. This accident-free introduction of a new technology is in itself a remarkable achievement and, so far as it goes, vindicates the emphasis on safety which has always been fundamental to nuclear power programs. However, the only conclusion that can be drawn from it is that any accidental release is likely to be less frequent than about 10^{-2} per reactor operating year. Using this maximum accident frequency figure, the risk to the individual at 1 km, estimated above as about 3×10^{-6}/yr following a release of 1000 Ci of ^{131}I, is reduced by the frequency of the accident to about 3×10^{-8}/yr. In reality, it is even less than this since in the event of the accident occurring the wind direction may not be in the direction of the nominated individual. Making allowance for this fact, the assessed risk rate to a specified individual becomes smaller than 10^{-9} per year for a release of 1000 Ci of ^{131}I.

In deriving this risk rate, a relatively small release of activity has been postulated and it has been demonstrated that, with an accident frequency of order 10^{-2}/yr, the risk to an individual is vanishingly small. If the postulated release was much greater (10^6 Ci, say), the risk, assuming a linear dose/risk relationship, would be correspondingly higher if the frequency of this large release were to remain as high as 10^{-2} per reactor year. The engineering of power reactors is certainly designed to prevent such large releases occurring at this frequency. If the true frequency of a large release lies below, say, 10^{-4} per reactor operating year, then experience is not going to establish this rate until reactors have accumulated one hundred times as much experience as at present, and if as is believed, the true accident rate for large releases is even lower than this, there will be no practical experience on which to base an estimate of risk within the foreseeable future.

The recognition of the limitation of the occurrence of accidents as a guide to the standards of safety in power reactors has led, from the earliest days, to a systematic analysis

of the potential causes of accidents, the course they may
follow, and the consequences which might ensue. If the risk
posed is to be assessed, it becomes necessary to assign probabilities to initiating faults and to consider what standards
of reliability may be required of the safeguards built into
the system to minimize the consequences. These matters are
considered in later chapters.

The problem, then, is to define an acceptable standard
of safety to which designers and safety assessors may work,
and which will result in acceptable standards of risk both to
individuals and to the community at large. One approach to
defining such a standard is to assess the costs resulting from
an accident and then to set a safety criterion on an economic
basis. The main costs envisaged following an accident would
result from the following.

(a) Fatalities resulting from the accident.
(b) The cost of milk restrictions where pasture is contaminated by iodine fallout.
(c) The need to evacuate people from areas suffering a
persistent increase in radiation level from the deposition of
the long-lived isotope cesium-137.

The probable number of cases of thyroid cancer to be expected
in the population around a reactor giving rise to an accidental
release can be calculated knowing the probability of different
weather conditions (which govern downwind dispersion), wind
directions, and, of course, the distribution of population
around the site. It is also necessary to know the relationship between dose and risk. A given release may occur in any
weather condition and with the wind blowing in any direction.
The extremes possible are the following.

(1) The best outcome, in which the release occurs in
the weather most favorable for dispersion and with the wind
blowing towards the least populated sector.
(2) The worst outcome, the release occurring when the
weather is least favorable for dispersion and the wind blows
towards the most highly built-up sector.

Between these two extremes will be a range of consequences of
various probabilities. Table V sets out the result of calculations of the probability of thyroid cancers in the population
for releases of 10^4, 10^5, and 10^6 Ci of ^{131}I for UK weather
conditions. Three types of site have been considered: a remote site with only a nearby village and scattered rural population for many miles; a semiurban site of the type currently
selected for the siting of advanced gas cooled reactors in
the UK adjacent to a developed area and with many towns in the

4. The Calculated Risk—A Safety Criterion

TABLE V

Probable Cases of Thyroid Cancer Arising from the Release of 10^4, 10^5, or 10^6 Ci of ^{131}I on Typical Sites

Release (Ci of ^{131}I)	Casualty range	Probability		
		Remote site	Semiurban site	City site
10^4	1-10	0.062	0.042	0.52
	10-100	0.0	0.023	0.35
	100-1000	0.0	0.002	0.035
	>1000	0.0	0.0	0.0
10^5	1-10	0.23	0.30	0.1
	10-100	0.094	0.23	0.60
	100-1000	0.005	0.055	0.28
	>1000	0.0	0.002	0.03
10^6	1-10	0.28	0.073	0.0
	10-100	0.21	0.29	0.0
	100-1000	0.091	0.200	0.19
	>1000	0.005	0.05	0.81[a]

[a] Obtained by difference.

middle distance; and a hypothetical site at the center of a city of 4×10^6 population.

It is similarly possible to calculate the probable distance of milk restrictions and long-term ground contamination from cesium-137 [9]. Costs can then be calculated if values are attached to the losses incurred. For purposes of illustration, somewhat arbitrary assumptions have been made but it turns out that the conclusions drawn are not much affected by quite wide variation in these suppositions.

The results are shown in Table VI, together with the suppositions made about cost. The distance ranges shown are the mean values for UK weather variations. It will be seen that the costs for modest releases are relatively trivial, but that there is a rapid rise in cost for accidents in which significant numbers of people may be affected by a persistent background level in excess of 0.5 R/yr from deposited cesium. For large accidents this appears to be the dominant cost, comprising some 80% of the total.

It is now possible to attempt a simple exercise to see, in purely economic terms, what limits these cost figures may impose on the acceptable frequency of accidents. A typical nuclear reactor of 600 MWe output generates power worth about

TABLE VI

Costs of Accidental Releases[a]

Size of release (Ci of ^{131}I)	Fatalities Thyroid	Fatalities Other	Total fatalities	Cost of fatalities (£)	Milk range (miles)	Cost of milk restrictions (£)
10^3	—	—	—	—	4.3	8,000
10^4	1	5	6	—	15	50,000
10^5	10	50	60	600,000	50	600,000
10^6				6,000,000	180	7,000,000

Cesium-137 range (miles)	Population in cesium-137 range	Loss of GNP (£)	Cost of rehabilitation (£)	Total costs
0.22	Few	5,000	25,000	4×10^4
0.78	200	200,000	1,000,000	10^6
2.8	34,000	3.4×10^7	1.7×10^8	2×10^8
10	100,000	10^8	5×10^8	6×10^8

[a] Cost assumptions imply: each fatality involving compensation of £100,000; the prohibition of milk consumption within the affected area compensated at current UK retail prices; and people evacuated in areas where the persistent gamma background exceeds the value of 0.5 R/yr recommended by the ICRP as a maximum for limited numbers of the general public (a loss of production of £1000 per year per person is roughly the mean UK gross national product).

4. The Calculated Risk—A Safety Criterion

£10^7 per annum at the station output. If the cost of insuring against accidents is not to add significantly to the cost, then a sum equal to, say, 1% of this amount could be set aside each year to meet the cost of an accident. With such a sum, large accidents costing about £10^9 could be accepted with a frequency of once in 10,000 operating years or 10^{-4}/yr. Smaller accidents could be accepted at a greater frequency.

A frequency of 10^{-4}/yr for a large accident is low and might be deemed acceptable if the total number of plants was small. If, as seems possible, the number of nuclear plants in the world approaches the number 500-1000 by the end of the century, then a major accident would statistically be expected every 10-20 years. In view of the dislocation to the lives of people and the public controversy about the acceptability of risks from nuclear power, it is doubtful if this purely economic argument sets a sufficiently low frequency for major accident events, taking a world view. Even though many "conventional" industries may offer greater risks to human life from major accidents, it is inevitable that the nuclear industry will be required to meet higher standards because of the nature of the risk and the more emotive reaction to radiation risks as compared to the better understood risks of other industrial technologies.

It is, therefore, essential to attempt to define a safety criterion for the nuclear industry which can be shown to be conservative in its effect and which reflects the need for low frequencies of occurrence for the larger accidents. The criterion should also reflect the understanding that reactors cannot be typified by a single value for release. All power reactors contain many safety systems and devices designed to prevent unsafe conditions arising, and to minimize the release to the atmosphere of any radioactive material. If the design intention is completely fulfilled, there will be no accidental releases of activity. Partial failure is countered by providing redundancy of essential systems. Nevertheless, if sufficient failures of equipment are postulated, unsafe conditions can arise and activity will be released within the reactor system and ultimately, some part of this, to the atmosphere. We should also accept that, if we are seeking to analyze events having low probabilities of occurrence, we must recognize the chance that the event has been wrongly foreseen so that the equipment, although perfect, is not able to prevent the development of events. With a sufficiently large postulated number of failures, or failures of a structurally catastrophic nature, very large releases can be envisaged. With this complexity it is feasible to imagine that if a large number of similar reactors were operated for a sufficiently long time, all modes of failure and all possible releases would be observed. This has indeed been the pattern followed in the past by most developing technologies and has led, by direct experience, to the development of rules and codes of practice and inspectorates to enforce suitable standards.

Since high standards have been set from the onset of the nuclear industry and it is therefore believed that large accidents will prove very infrequent, the derivation of the frequency spectrum of accidents by direct observation as described is not practical. The alternative is to derive safety characteristics by analysis of the behavior of reactor systems, the analysis being supported in its various aspects by an extensive research and development program. Such a program gives confidence in the basic data used in fault analysis and provides guidance on the uncertainties involved. When all this has been undertaken and the behavior of the plant assessed, there remains the question of whether a satisfactory standard of safety has been achieved for all postulated accidents. There is also clearly a need to assess the priority to be accorded to different accidents having different potential accident consequences, so that the best overall safety may be achieved.

One way of quantifying these questions has been suggested by Farmer [11], who pointed out that because of their proportionately greater impact large accidental releases should be possible only at a much lower frequency than smaller accidental releases. This can be done by defining an "acceptable" frequency of release for accidents of different magnitude and then by imposing safeguards on the plant ensure that this standard is achieved. A form of this criterion which has been used within the UKAEA for the assessment of prototype reactors is shown in Fig. 1. The ordinate of this curve represents the frequency of accidents, expressed in terms of the probability per reactor year of operation, and the abscissa represents the size of activity released, expressed for thermal reactors in terms of the ^{131}I content of the released fission products. Both axes are on a logarithmic scale.

Over much of the potential release range there is a negative slope of -1 on this logarithmic scale, and the criterion may be interpreted in the following way to a fair degree of approximation. Releases of a magnitude of 10^6 Ci or greater of ^{131}I should not exceed a frequency of 10^{-6}/yr of operation, releases of 10^5 Ci or greater should not be more frequent than 10^{-5} per operating year, and so on. At low releases there is a limit to the frequency of releases, more related to the undesirability of frequent small accidents than to the real consequences which may ensue. The quantitative interpretation of this type of criterion has been defined and explored in Ref. [12].

4. The Calculated Risk—A Safety Criterion

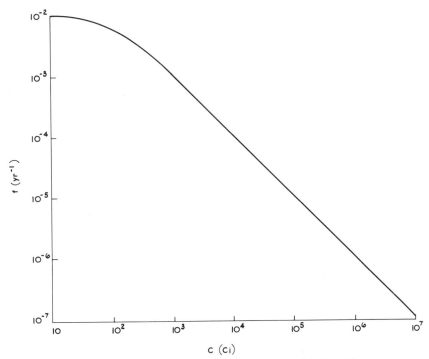

FIG.1 *Accident release frequency limit line.*

The acceptance of some such criterion as described above immediately imposes the discipline of quantitative assessment upon both the reactor designer and the safety assessor. This is not to imply that only if a criterion is stated numerically will the safety argument be conducted on some quantitative basis, but it does emphasize the virtue of plant analysis in such a way. The designer is conscious of the fact that he will be required to justify his chosen scheme or system by arguments showing that the necessary defined standard is attained, and that the safety assessor will be required to be convinced of the validity of this numerical justification. There is a clear impetus to understand the performance of subsystems in the safety chain better and to initiate research into plant analysis methods and into the collection of failure data of components and plants.

Having now defined safety requirements in a quantitative way, it is possible to assess the risk which would be imposed upon individuals and the community by a reactor which conformed exactly with the defined standard. The averaged risk to the individual may readily be calculated if we assume him to be exposed for a sufficiently long time to a reactor giving rise to accidental releases in the way specified as acceptable by

the safety criterion (Fig.1). If the individual lives at a
distance of 1 km from the reactor, the wind frequency in his
direction is 10% and the risk of contracting thyroid cancer
is taken to be 3×10^{-5}/rad, then in average weather conditions the risk of thyroid cancer is about 3×10^{-8}/Ci ^{131}I
released, and the risk of death is about 3×10^{-9}/Ci released.
Taking account of other cancers, the risk is somewhat higher
but not more than about 10^{-8}/Ci of iodine in the released mixture. A single accident, represented by a point on the control
curve (Fig.1), corresponds to a mean release of 1 Ci/yr, and
the total curve corresponds to a mean release of a few curies
per year taking all possible accidents into account. It follows, therefore, that a reactor operating in conformity with
the criterion implies a risk rate of between 10^{-7}/yr and 10^{-8}/yr
to the nominated individual at 1 km. Of course, this is a somewhat simplified view, since clearly no risk arises until an
accident has actually occurred, but it has its use in enabling
a statistical risk arising from a possible reactor accident to
be compared with the other risks of life discussed earlier
(see Table I). For people living at a greater distance, the
risk will be smaller reducing approximately as the inverse
square of the distance. Emergency schemes at reactor sites
provide additional protection for the nearby population through
evacuation measures and the issue of stable iodide tablets to
block the uptake of iodine by the thyroid gland.

It will be seen that if the control curve can be worked
too, the nuclear risk is certainly small compared with other
risks and could be judged adequately low. Vinck [13] has
pointed out that risk rates of about 10^{-3}/yr to the individual
are usually considered of such importance that steps are taken
to reduce the risk if possible. Lower-order risks are treated
with proportionately less concern and risk rates of 10^{-6} or
less per year do not cause much concern to the population.
This accords with the general assessment of the more hazardous
occupations (Table III), where the lifetime risk amounts to
about 5%, giving for a 40-yr working life a risk rate of about
10^{-3}/yr, compared with a rate of about 10^{-4}/yr in all manufacturing industries. Such risks are, it may be suggested, undertaken knowingly and voluntarily, whereas the risk from a nearby
nuclear station is one that is imposed and should therefore be
of much lower order. A study of the difference between "voluntary" risks has been made by Starr [14]. A risk rate in the
range 10^{-7}-10^{-8}/yr may be considered adequate to meet the requirement for low-risk "involuntary" exposure.

Having demonstrated that the individual risk is, at least
arguably, adequately low, it is necessary to consider what may
be termed the community risk, that is, the integrated risk to
all individuals in the surrounding community. It is almost

4. The Calculated Risk—A Safety Criterion 63

commonplace that society reacts strongly to events causing large numbers of deaths or disruption to normal life while tolerating events leading to small numbers of casualties. For example, aircraft crashes involving large numbers invariably rate headlines in the news media, although the total passengers killed in this way on scheduled flights in the UK averages only about 1% of those killed in road traffic accidents in recent years. In considering nuclear accidents, therefore, it is important not only to estimate the averaged risk to the individual, but also to establish the frequency with which accidents producing a varying number of casualties may occur.

As already mentioned, for small releases the important consequences following a release of volatile fission products result from the iodine component. The immediate and unavoidable hazard is inhalation from the cloud giving rise to a radiation dose to the thyroid gland. As the size of the release increases, the external whole-body dose from the cloud also increases and, for sufficiently large releases (containing around 10^6 Ci of iodine-131) may reach values at which short-term effects (nausea, sickness, etc.) may become important. The semilethal dose is unlikely to be reached where population density is high (i.e., beyond distances of about 500 yards from the reactor), since in short-term releases as the size increases beyond 10^6 Ci of ^{131}I, the self-heating of the fission products becomes of importance and causes the plume of activity to rise, reducing doses to people on the ground by about a factor of ten. If the release is prolonged, natural radioactive decay of the shorter-lived fission products will again reduce the total external dose to people downwind.

The consequence of these considerations is that, even in large releases, and taking population doses over many miles into account, the potential deaths resulting from thyroid cancer will still be dominant, and so we may continue to use the incidence of thyroid cancer as a good measure of the accident potential over the entire range of possible accidents. If the iodine content were filtered, as is the intention in many accident situations, these conclusions would obviously require modification, but the total risk would be considerably less than an unfiltered release. The problem of estimating the probable consequences of accidental releases may be considered in the following way. Suppose that the size of the release is defined; for example, let us suppose that a release of 10^4 Ci of ^{131}I has actually occurred. The dose received by individuals at a given distance downwind will depend upon the weather at the time of release and may, in the worst weather, be over 100 times as great as in the most favorable conditions.

The probability that the dose is within given limits depends upon the probability of the weather dispersion conditions also lying within the appropriate limits. Thus, knowing the probability distribution of dispersion conditions, the probability distribution of dose to these individuals may be derived. Weather probabilities may be derived from meteorological data. The true probability distribution of dose requires that we assign a probability to the release of 10^4 Ci considered, and also to the probability that the wind blows towards the group of people at the given distance. Knowing the relationship between dose received and the risk of contracting thyroid cancer, it is then easily possible to transform the probability distribution of dose to a probability distribution of thyroid cancers. The calculation must be repeated for all population groups around the site and the results added to obtain the risk distribution for the whole community.

Table V showed the result of such a calculation for releases of 10^4, 10^5, and 10^6 Ci of ^{131}I on three types of site. In all cases the true probability of those events may be obtained by multiplying by the probability of the stated release actually occurring. Thus, if it is believed that the probability of a release of 10^4 Ci occurring is 10^{-4}/yr, then the probability of 1-10 cases of thyroid cancer being caused is 4.2×10^{-6}/yr on a semiurban site.

The spread of possibilities shown in Table V has arisen because, for a specified release, account has been taken of both the weather and wind direction probabilities. If we wish to estimate the total spread of effects arising from a reactor having an accident spectrum of releases conforming with the type of safety criterion shown in Fig.1, it is necessary to include this distribution of releases in the calculation. In effect, three probability distributions must be combined.

    1. The probability distribution of accidental releases (defined, say, by the safety criterion of Fig.1 or by a detailed plant analysis).
    2. The probability distribution of weather conditions (derived from meteorological data).
    3. The distribution of wind direction, again obtained from meteorological information.

The calculation is an extension of that described above for specified releases, but now the whole accident criterion is represented by choosing suitable pairs of releases and release frequency throughout the range of release and adding all results together. In this way any release criterion can be represented to the desired standard of accuracy. All possible combinations of circumstances are thus included, from the

4. The Calculated Risk—A Safety Criterion

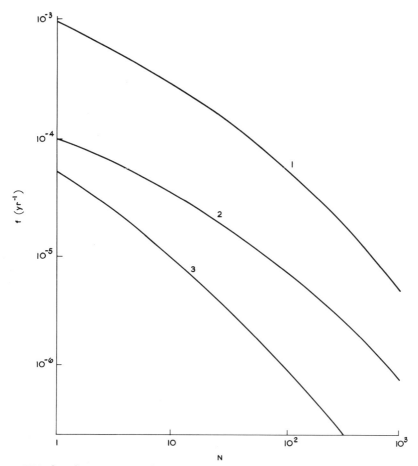

FIG.2   f-N curves for three population distributions: 1, hypothetical city; 2, semiurban locality; 3, remote region.

smallest release in the best weather and with the wind blowing in the most favorable direction, to the largest release in the most adverse weather and with the wind blowing towards the most heavily populated sector. The result is represented as a frequency distribution of the occurrence of cases of thyroid cancers throughout the population surrounding the site.

Figure 2 shows the curves obtained for the three typical sites referred to earlier, and for which Table V shows the distribution of thyroid cancers following specified releases of 10^4, 10^5, and 10^6 Ci of ^{131}I. Suitable integration of these curves would enable a similar table to be drawn up, but approximate values can be obtained directly from the curves in the following way. If the frequency corresponding to a particular number of thyroid cancer cases is read off the curves, this will represent the frequency with which cases of cancer within the decade interval about the chosen number will occur. For example, considering the line representing the remote site, the frequency with which 100 cancer cases is to be expected is about 10^{-6}/yr. This may be interpreted to mean that the probability of accidents causing cancers in the decade interval between 30 and 300 is about 10^{-6}/yr, or once in a million reactor operating years.

Since the sites considered span the spectrum of practical possibility in a densely populated country such as the UK, the difference between them represents the maximum advantage which can be derived from site selection. It will be seen that in the region of most concern (events causing thyroid cancers in the range 10^2-10^3) there is about a factor of 10 between each of the curves. Choosing semiurban rather than rural sites thus increases the frequency of given consequences by about a factor of ten, and a further factor of ten is lost if city-center siting is chosen. This result occasions no surprise since the relative mean population density between the remote site and the city site is about 1:100. It is, though, pertinent to observe that although siting can effect at most a factor of 10^2 in risk, the range of iodine-131 inventory in a reactor exceeds seven decades. It is apparent that the safety of a nuclear power reactor resides essentially in the safety characteristics of the plant itself, and that siting can make only a relatively small contribution to overall safety. Problems arising following an accident will, of course, be more amenable if population in the immediate vicinity of the station can be dealt with by an effective emergency scheme, particularly as the release of activity in accidents is likely to be prolonged rather than sudden. Details of the methods developed for risk evaluation which have been described here in a qualitative fashion are given in [11,12,14].

Having reached this stage in risk evaluation, it is worth returning to the basic purpose of safety studies and application in the nuclear field. The object of all such work is the protection of the individual and the community from the effects of ionizing radiation. No absolute guarantee of complete safety can be offered, but the nuclear industry, conscious from its beginnings of the public safety aspects, has perhaps more than any other industry devoted considerable effort, not to responding to accidents after the event, but to studying the behavior of its plant in possible but unlikely eventualities. Given a safety criterion relating the frequency and size of accidents, and a knowledge of the variation of meteorological conditions, it is possible to evaluate the probability of dose distribution in the surrounding population. If the dose/risk relationship is known (or postulated), the frequency

4. The Calculated Risk—A Safety Criterion

distribution of casualties (i.e., cases of cancer) can be calculated, with an accuracy as good as the basic data employed. Since all reactors are capable of accidental releases, the design of a reactor and its siting together inevitably entail a statistical risk of the sort described for the surrounding area. If the release characteristics of the reactor can be deduced by assessment, or if an acceptable safety criterion can be defined, the corresponding risk can be calculated.

For simplicity of exposition the argument in this paper has been presented primarily in terms of the risk from the induction of thyroid cancer resulting from radioactive iodine released in an accident. The isotopic composition released in any accident strongly depends upon the course of events in the accident sequence and the effectiveness of plate-out, partition, or filtration in depleting particular components from the mixture released directly from the fuel. More sophisticated analysis accounts for the appropriate mixture of isotopes released, which may also be determined to some extent by the type of nuclear reactor under consideration. The sodium coolant of a fast reactor, for example, is a good medium for the absorption of iodine so that the spectrum of isotopes released from such reactors involving fuel melting may be quite different from that arising in light water-cooled reactors. The choice of one isotope (iodine) on which to base a safety criterion may, therefore, be too great a simplification, although useful for many thermal reactors. One approach could be to compare the risk from each isotope with that from iodine, and so to deduce criteria for a range of fission products. Such a procedure is clearly cumbersome and could lead to a proliferation of standards. Moreover, it would tend to obscure the prime purpose of safety evaluation, which is to assure sufficient protection of the population. The real need is to define a standard of adequate and acceptable protection from which secondary criteria, of the type shown in Fig.1, relating the frequency, size, and nature of release can be derived.

The risk to the individual has been estimated as in the range 10^{-7}-10^{-8}/yr and can, for the safety criterion considered here, be compared directly with other risks to which people are exposed, and conclusions may then be drawn about the acceptability of such risk levels. It is more difficult to provide guidance on the comparison between the community risk arising from reactors and other risks, either from natural causes or from human activities. There is probably little profit to be gained from assessing the frequency and size of natural disasters, such as storms and earthquakes, since these tend to recur in particular areas. It is interesting to note, however, that while some populations have perhaps little choice but to endure the hardships and risks of such events, there

are many instances of people in developed countries continuing
to live in earthquake areas or in the vicinity of volcanoes
although historically the frequency of major disasters lies in
the range 10^{-2}-10^{-3}/yr.

A natural background risk which is quite random arises
from the bombardment of the earth by meteorites, some of which
are sufficiently large to cause widespread devastation, as in
the crater in Arizona. Astronomers have calculated the frequency and size distribution of meteorites incident on the
earth [16], and the resulting distribution of deaths can be
obtained from a knowledge of population density. Figure 3
shows the result of such a calculation [17] for England and
Wales, plotted together with the risk curve of 30 reactors on
semiurban sites. This latter is derived from Fig.2 by multiplying the ordinate by 30 and strictly represents cases of
thyroid cancer, rather than death. Taking account of other
cancers arising from the irradiation of other organs, it may,
however, be regarded as an approximate estimate of the distribution of deaths resulting from induced cancers from reactor accidents, subject of course to the explanation given.

Manmade accidents spanning a wide range of casualties
for comparison with nuclear risks are themselves also infrequent, although fires and explosions have in the past caused
casualties to exceed the hundred in individual cases. For
instance, numerous explosions have occurred from leakage of
natural or petroleum gases, such as the liquified natural gas
explosion in Cleveland, killing some 130 people. Recent examples of fires are those at the "Summerland" pleasure building in the Isle of Man (UK), causing 50 deaths, and the fire
in a Tokyo store killing 66, both events occurring in 1973.
There are also many aircraft crashes involving large numbers
of people in single events, but it may be argued that this is
a risk voluntarily undertaken and not truly random over a population not all of whom choose to fly.

Nevertheless, the population on the ground are subject
to a risk should an aircraft crash, and the number killed on
the ground could be substantial should the aircraft fall in
a densely populated area. If only random crashes are considered (this excludes those crashes associated with take-off
and landing), the distribution of numbers killed will depend
upon the population density distribution on the ground. A
distribution of this type, based upon one such random crash
per year in England and Wales, is shown in Fig.3. The casualties here refer to people killed on the ground; the passengers
in the aircraft are not included. In this figure there are
thus three distributions of risk to the community—from a
natural background risk of low order (meteorites), from a manmade risk from overflying aircraft, and from 30 reactors, each

4. The Calculated Risk—A Safety Criterion

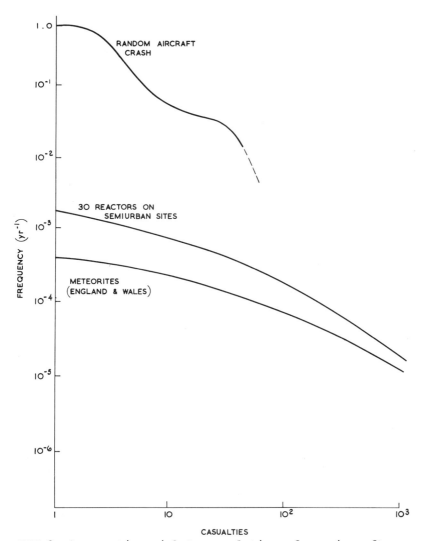

FIG.3 Comparative risk to populations from aircraft, meteorites, and nuclear reactors.

conforming to the safety criterion of Fig.1. All are hypothetical in the sense that all have been derived from theoretical analysis rather than from an assessment of past experience. This is perhaps an indication that the techniques of safety evaluation of events of low probability, developed in the nuclear industry, could be profitably applied to low-probability/high-consequence events in more conventional industry, where the potential for disaster may not have been so exhaustively analyzed.

The community risk presented by nuclear reactors can therefore be estimated if the reactor characteristics or a safety criterion are specified in terms of the frequency of accidents as a function of their magnitude. It can be compared with other man-imposed risks or with aspects of the natural background. The comparison, shown in Fig.3, is then a guide to the right choice of safety criterion. As can be seen there, the risk to the community of a network of reactors conforming to one suggested criterion lies close to a remote background risk which occasions little alarm to most people. The nuclear risk also appears to be very comfortably less than the risk we accept from commercial air traffic, and there may well be risks from other human activities, such as the use of combustible gases or in chemical operations, which because of expansion of scale now present risks to society worthy of closer analysis.

It should, perhaps, be pointed out that the licensing of reactors for power generation has not, so far, been based upon an explicit quantitative criterion of the sort shown in Fig.1, but rather upon a detailed examination of possible fault sequences defined in a more deterministic fashion. Judgment of the acceptability of designs has then been made on the basis of good engineering practice and an understanding of the physics, chemistry, and kinetics of the system. In the US, a study has been commissioned by the USAEC to examine, by the application of probability and reliability techniques, two operating reactors, one a pressurized water reactor and the other a boiling water reactor. The study will result in the assessment of risk to the surrounding population from the range of accident sequences identified. Preliminary findings, published in draft form, indicate that for these reactors the risks to individuals and to the community are much less than those presented by natural events such as floods or hurricanes or by other manmade technological risks. The results of this study of two operating reactors are, thus, in general accord with the conclusions described in this chapter based upon an idealized reactor conforming with a defined safety criterion.

Nevertheless, when all this has been said, there are intangibles involving the more emotive nature of the risk from radioactive substances and the long time scale over which the cost of accidents may be paid, which favor the adoption of safety standards more restrictive than perhaps pure logic would dictate. Certainly a serious accident to a nuclear reactor in the near future anywhere in the world would have repercussions on the building program in all countries and could create the antipathy to nuclear matters inevitable after the initial military use of fission power. Since that time public anxiety has focused successively on the genetic risk, the levels of

4. The Calculated Risk—A Safety Criterion

"permissible" dosage, the possibility of catastrophic accidents, and the long-term storage of fission products. With the development of fast reactors, the issue of plutonium safeguards is coming to the fore. The risks presented by nuclear power have been more thoroughly explored in public debate than any other aspect of industrial development, and it is only right that the industry should be required to allay public fears of this kind by presenting the best assessment of the various risks and by attempting to put them into the context of the general background of risk to which human society is exposed. The choice, ultimately, is for the public to make. The benefits of nuclear power are clear and are becoming more and more apparent as shortages of fossil fuel develop. The risks of nuclear power have been explored, analyzed, presented and discussed, and a background of substantial operational experience has been built up over the past twenty-five years. It will be important to ensure that safety remains in the foreground of nuclear development and that complacency is not permitted to develop. The spread of nuclear power will provide an impetus towards internationally accepted safety standards which are not yet in sight for power reactors although they have been developed, under the stress of necessity, to facilitate the international transport of radioactive materials.

References

1. Annual Abstract of Statistics, No. 109, Central Statistical Office, HMSO, London (1972).
2. US National Safety Council. Accident Facts (1972).
3. US National Academy of Sciences and National Research Council. The Effect on Populations of Exposure to Low Levels of Ionizing Radiation, Report of the Advisory Committee on the Biological Effects of Ionizing Radiation (November 1972).
4. Environmental Protection Agency. Estimates of Ionizing Radiation Doses in the United States, 1960-2000, Special Studies Group, Office of Radiation Programs, Rockville, Maryland (1972).
5. Sowby, F. D. Some risks of modern life, IAEA Symposium on Environmental Aspects of Nuclear Power Stations, New York, August 10-14, 1970, Paper No. IAEA-SM-146/48.
6. Hill, Sir John. The Abuse of Nuclear Power, *ATOM* No. 239, UKAEA (September 1976).
7. United Nations. Ionizing Radiation: Levels and Effects, A Report of the United Nations Scientific Committee on the Effects of Atomic Radiation, New York (1972).

8. Dolphin, G. W., and Marley, W. G. Risk Evaluation in
    Relation to the Protection of the Public in the Event of
    Accidents at Nuclear Installations, UKAEA Rep. AHSB(RP)R93
    (1969).
9. Beattie, J. R., and Bell, G. D. A Possible Standard of
    Risk for Large Accidental Releases, IAEA Symposium on
    Principles and Standards of Reactor Safety, Jülich, 1973,
    Paper No. IAEA-SM-169/33.
10. Farmer, F. R. IAEA Symposium on Principles and Standards
    of Reactor Safety, Jülich, 1973, invited paper.
11. Farmer, F. R. Siting Criteria - A New Approach, IAEA
    Conference on Containment and Siting, Vienna, April 3-7,
    1967, Paper No. SM-89/34.
12. Beattie, J. R., Bell, G. D., and Edwards, J. E. Methods
    for the Evaluation of Risk, UKAEA Rep. AHSB(S)R159 (1969).
13. Vinck, W. Nuclear Power Generation in Western Europe,
    Safety and Environmental Implications, paper presented
    at the International American Nuclear Society Meeting,
    Washington, November 12-16, 1972.
14. Bell, G. D. Risk Evaluation for Any Curie Release Spec-
    trum and Any Dose-Risk Relationship, UKAEA Rep. AHSB(S)
    R192 (1971).
15. Cagnetti, P., Frittelli, L., and Nardi, A. Some Consid-
    erations on the Reliability Function of a Nuclear Plant
    with Respect to its Site, RT. PROT-(72)19 (1970).
16. Blake, V. E. A Prediction of the Hazards from the Random
    Impact of Meteorites on the Earth's Surface, Sandia Lab-
    oratories, New Mexico, SC-RR-68-838 (December 1968).
17. USAEC. Reactor Safety Study (draft), WASH-1400 (1974).

5

Quantitative Approach to Reliability of Control and Instrumentation Systems

A. Aitken

Introduction

In this chapter some of the safety aspects associated with various control and instrumentation schemes for nuclear reactors are reviewed. Such schemes are particularly amenable to reliability analysis in quantitative terms, partly because reliability data can be more easily obtained, but mainly because in such systems enhancement of reliability can be readily achieved by application of "redundancy" and "diversity." These terms will be explained, a typical automatic protective system defined, and methods of analysis and assessment outlined, but first some basic ideas on quantitative reliability are given.

Reliability and Probability

The reliability of protective equipment is especially important when such equipment is used in the protective circuits of large plants or systems. Typically, important operational aspects are: (a) availability, and (b) safety. In any system, the equipment availability should be high for maximum system effectiveness. Unscheduled shutdowns due to equipment failures can be costly. For example, if a nuclear reactor in a generating station is inadvertently tripped for a period of several hours, then the loss in electricity sales amounts to several thousand pounds. As a converse requirement, safety demands that the plant must be shut down should dangerous conditions begin to arise. This is necessary in order to guard against possible damage to costly plant equipment and possible hazards to the public and the operating personnel. These two conflicting requirements (availability and safety) can be considered under the term "reliability." A definition of reliability is as follows:

> The reliability of an item is the probability that that item will perform in the manner desired for a specified period of time under the specified environmental conditions.

The word "item" can cover a wide range in its use; for example, it could mean a single amplifier or a large system.

In reliability work individual probabilities of events can very often be expressed as functions of time. A common probability function of this type is that known as the exponential distribution which may be expressed by the relationship

$$p(t) = 1-e^{-\theta t} \qquad (1)$$

where t is the time, θ the failure rate of the item in faults per unit time, and p(t) the probability that the item will have failed by the time t when it was working at time zero. This formula can be expanded using the exponential series to show that when θt is much less than unity, then

$$p(t) \simeq \theta t$$

By way of example, if an item had a failure rate of 0.1 f/yr (faults per year) and the time of interest t was 3 months (0.25 yr), then

$$p(t) = 1-e^{-0.025} \simeq 0.025$$

Thus, there is a 1 in 40 chance that at the end of a 3-month period, the item will be found in the failed state when it was operating correctly at the beginning of the period.

Further treatment of this and other probability distributions may be found in Ref.[1], which also contains an extensive bibliography on the subject. Particular application to the reliability analysis of automatic protective systems is contained in Refs.[2] and [3].

Equipment Failure Rates

Experience suggests that most equipments conform, on average, to a fairly standard type of failure pattern with respect to time. A typical pattern of this sort is illustrated in Fig.1. This is a plot of average failure rate

5. Quantitative Approach to Control and Instrumentation Systems

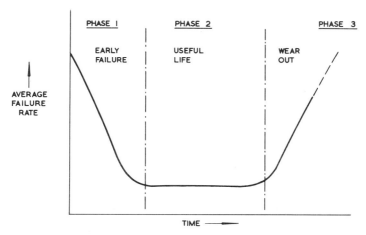

FIG.1 Failure rate characteristic.

against time and is divided into three phases. The first phase represents a failure pattern which arises from initial production, test or assembly faults. The last phase illustrates the effects of aging when the equipment is beginning to wear out. In between is a phase which may be termed the "useful life" where the average failure rate remains sensibly constant at some particular value. The assumption of a constant failure rate can be shown to lead to a probability of failure which follows the exponential distribution previously quoted in Eq.(1).

Most equipments are designed and used so that they operate in phase two, i.e., the useful life phase. This is normally achieved by the application of soak testing, commissioning procedures, and the weeding out of early failures. Arrangements are also generally made to withdraw equipment or to carry out service so that it does not enter the wear-out phase.

There are three main ways in which the average failure rate may be obtained: (i) field experience, (ii) sample testing, and (iii) prediction. Generally, with new or relatively new equipment there is not enough information available from field experience or sample testing, and recourse has to be made to some form of theoretical prediction of failure rate. This is especially the case for equipment of high reliability where the rate of occurrence of failures is low and practical failure rate figures are unlikely to become available for a relatively long time. Where the fail/dangerous rate of an equipment is very low (for example, 0.01 f/yr), a practical figure for this rate may never be gathered before the end of the equipment's life.

TABLE I

Average Failure Rates for Electronic Components[a]

Type of component	Assumed failure rate[b] (faults/10^6 hr)
Resistors (fixed)	
High stability, carbon	0.5
Wirewound	0.5
Composition, grade 2	0.1
Metal film	0.1
Oxide film	0.5
Capacitors (fixed)	
Paper	1.0
Metallized paper	0.5
Mica	0.3
Ceramic	0.1
Polystyrene	0.1
Valves	
Diodes	12.0
Double diodes	16.0
Triodes	19.0
Double triodes	28.0
Tetrodes	21.0
Pentodes	23.0
Transistors	
Germanium high power	1.0
Silicon high power	0.5
Silicon low power	0.05
Semiconductor diodes	
Germanium, point contact	0.5
Germanium, alloy junction	0.2
Germanium, gold bonded	0.2
Silicon, zener	0.1
Silicon, high power rectifier	0.5
Silicon, low power diode	0.02

[a] As assumed to be applicable to electronic equipment used in the automatic protective systems of land-based nuclear installations.

[b] These figures are average values applicable to large samples of components working at nominal ratings and operating under normal environmental conditions. A fuller selection of component values together with precautions to be taken in their use may be found in Ref. [1].

5. Quantitative Approach to Control and Instrumentation Systems

The theoretical prediction is based on the concept that even though a complete equipment may be of a relatively new design and manufacture, it is quite likely that the majority of its constituent parts will be made of components with well tried and proved performances. Information on the failure rates of components is generally far more abundant than that on complete equipments. This is particularly true in the electronic field. In most cases, equipments can be analyzed part by part in order to determine the effects on overall performance of the various modes of failure possible in each of the individual components. It then follows that, with a knowledge of the component failure rates, an overall equipment failure rate can be predicted. There are a number of reservations to this approach but with due care and attention quite accurate results can be achieved.

Many authorities have issued component failure rate figures over the past few years and, in particular, there has been a fairly large coverage of the electronic type of components. Prediction experience in the field of equipment used in nuclear installations has led to a set of assumed figures for the purposes of safety evaluation which have been derived from a number of available sources. A selection of these assumed figures is given in Table I by way of example. The figures are average values taken from large samples of the appropriate components over a number of years under normal operating conditions. Bearing this in mind, they should be used with care in particular or individual cases.

Typical Systems

Instrumentation provided to monitor the state of a reactor generally takes the form shown in the block diagram of Fig.2. Information about the reactor state is used in three ways. First, it is processed in the automatic control circuits and passed via the normal control elements to automatically maintain the reactor in a predetermined state. Second, it is passed via the automatic protective circuits and safety control elements to establish automatic control in the protective or safety sense. Third, the information is displayed to an operator who can then act to control the reactor manually.

The general instrumentation facilities, as can be seen from Fig.2, form the basis for all subsequent controlling functions. The instrumentation involves the measurement of various reactor parameters such as temperature, neutron flux, and coolant flow by appropriate sensors. These sensors and their associated equipment convert the information from the reactor into

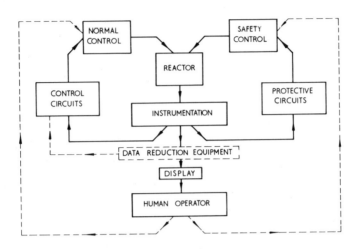

FIG.2 Reactor control and instrumentation.

suitable signals for feeding the control and display facilities. In a power reactor system the number of measurements required, typically two or three thousand, presents problems in handling and presentation. It becomes advantageous to use data reduction equipment (DRE). The core of this equipment is a digital computer which allows the operator to select the information he requires according to preset programs and also affords facilities for automatic logging of plant variables.

The automatic safety control—or automatic protective system—is the principal safety feature against the majority of accident conditions. It may be defined generally as all the equipment which automatically senses and produces action to initiate shutdown should a reactor fault condition arise. The rest of this chapter deals with analysis of relevant aspects of these systems and their equipments such as their reliability, their maintainability, and methods for assessing their safety. Some consideration is also given to the roles of the operator and the DRE (computer).

Analysis of Protective Systems

Protective Requirement

The possibility that faults might develop in a reactor and its associated equipment introduces a need to provide adequate protection against these conditions. For the various fault conditions the form that the protection should take and its objectives need to be clearly defined. The objectives

5. Quantitative Approach to Control and Instrumentation Systems

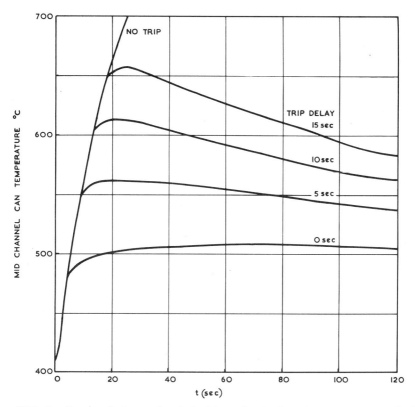

FIG.3 *Typical transient bottom duct fracture, gas-cooled reactor. Stepped in mass flow from 100% to 1%.*

may be at various levels, for example, if it is foreseen that various types of minor faults could occur fairly frequently, then the protective system might be designed to prevent any damage to the reactor fuel. On the other hand, if a fault is thought to be highly improbable, the protective system might be designed to prevent excessive escape of fission products from the reactor system or its containment. As a simple example, Fig.3 shows the way in which the fuel can temperature will change following a particular form of primary circuit failure on a graphite moderated gas-cooled reactor. The rise in temperature which occurs initially is arrested by shutting down the reactor so that the temperature reached by the fuel can depends upon the time delay in the protective system. Clearly, from this example the time dependence of all the factors which are relevant in this form of accident need to be carefully investigated and understood.

In order to achieve an adequate automatic protective system, it is necessary to have a definite knowledge of the

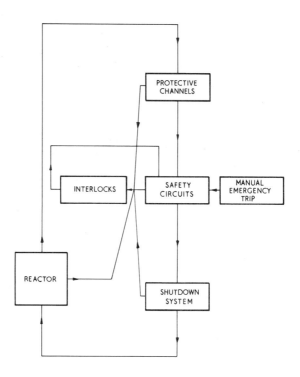

FIG.4 Block diagram of a reactor protective system.

reactor properties which, if monitored and maintained within limits, can be accepted as leading to the overall basic objective of reactor protection being fulfilled. For any system design it is obviously necessary for these relationships to be known in terms of quantities which can be measured in a practical fashion. Such considerations initially decide the information which is required to be derived from the reactor and fed into the protective system.

A reactor protective system may consist of several components, as shown in Fig.4, including the possible intervention of the operator through a manual emergency trip. The main components are identifiable as the "protective channels," the "safety circuits," and the "shutdown system." Performance criteria are required for each of these components defining the required response, accuracy, and reliability.

One final aspect of the required performance is generally taken into account. It is not normally sufficient that the protective system be capable of performing in the required way, but it is also important that this standard be met at all times and under all relevant environmental conditions. Careful cognizance is therefore taken of such conditions as changes in

5. Quantitative Approach to Control and Instrumentation Systems

environment, variations in power supplies, and operational procedures both under normal and abnormal reactor conditions.

More detailed aspects of the way in which the protective requirement may be met are now considered under the following headings: protective channels, safety circuits, and shutdown system.

Protective Channels

For every relevant reactor fault condition, the protective system is designed to be "diverse" from the common fault point of view. In some systems this is achieved by providing more than one method of detection and action.

It is normal at least to duplicate the measurement associated with any one parameter. A temperature measurement at one point in the reactor may be carried out by the use of duplicate thermocouples, duplicate cables, and duplicate temperature trip amplifiers. This duplication, or more generally the provision of more equipment than is required to carry out the measurement or control, is frequently described as the provision of redundant equipment so that even if a limited number of failures occur there is still adequate information or means for controlling the reactor safely.

The reliability of the system is normally affected by two main types of equipment faults. These are usually designated as "fail-to-danger" and "fail-safe." The former describes an equipment fault which would inhibit or delay automatic protective action. The latter describes an equipment fault which, irrespective of input, causes the protective system to move nearer to the trip point or to initiate trip action. Both types of fault have adverse effects. The fail-to-danger fault has a direct and detrimental effect on safety. The fail-safe fault may lead to an undesirable frequency of shutdowns which in a power producing reactor could incur heavy economic penalties. The fail-safe fault also has an indirect effect on basic safety in that it may bring the automatic protective system into disrepute.

Considerations of fail-to-danger and fail-safe equipment faults generally lead to a compromise on redundancy techniques. This compromise usually gives rise to a rejection of the simple duplicated or even triplicated system and replaces it by a number of redundant channels working on a majority vote principle. The simplest majority vote scheme requires three redundant channels from which the coincident output of any two are taken as truth. It should be emphasized that this "2-out-of-3 system" is not quite so reliable in dealing with fail-to-danger faults as the simple duplicated system, but it shows a considerable improvement with regard to the adverse effects of fail-safe faults. Normally, good equipment design against the fail-to-danger modes may make the 2-out-of-3 system quite adequate in the overall safety sense.

Hence, the equipment used in the protective channels is designed not only to meet the required dynamic performance specification but also to achieve a high standard of reliability with a minimum fail-to-danger fault rate. Minimizing fault rates of this class extends, of course, to the complete system concept apart from the individual items of equipment. As a simple example, loss of the polarizing voltage to an ionization chamber would normally be a fail-to-danger fault; however, this can be overcome by providing an additional trip signal if or when this supply fails. In general, loss of instrumentation supplies are designed to trip the particular protective channel concerned.

Safety Circuits

The need to achieve the required reliability standards leads to the adoption of redundant lines of equipment in any one protective channel, and the further requirement to process the redundant information so obtained on a coincident or majority basis. The form of the logic in processing the information depends upon the required degree of reliability and type of maintenance and service which is applied to each protective channel.

The logic employed is generally on at least a 1-out-of-2 basis plus the availability of adequate proof test facilities. Coincident or majority systems may be used but they normally still meet the 1-out-of-2 criterion. For example, under test conditions a 2-out-of-3 system is not deliberately moved to a 2-out-of-2 condition unless, of course, the diversity and redundancy of equipment required to protect against any incident is adequate for the purpose under the conditions of test.

Early, and many current, reactor safety circuits employ relay networks to carry out the required logic functions. Figure 5 shows a single guard line of this type. Two methods of operation are illustrated, i.e., 1-out-of-2 and 2-out-of-3. These give the logic A or B in the first case, and (C and D) or (D and E) or (C and E) in the second case. It should be noted that an additional advantage of the 2-out-of-3 method of working is that it allows on-load testing and maintenance on one instrument line while still maintaining an adequate standard of overall safety. The fail-safe principle is illustrated by operating the relays in the normally energized state. An earth fault on the guard line would cause the line to move its state to a trip condition depending upon the relay coil

5. Quantitative Approach to Control and Instrumentation Systems

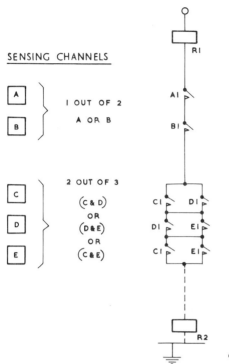

FIG.5 A single guard line employing relays.

resistance and earth fault resistance. If the line is not tripped, it could still be opened by a normal initiating signal. This is an aspect of a fail-safe guard line design employing solid earthing of one side of the supply and which uses "double-ended" monitoring relays.

Usually the safety circuits represent the focal point of the whole protective system, and it is essential to minimize the effects of faults at this stage. Ideally, any common point of the system should occur functionally as late in the system as possible, thereby maintaining the maximum safeguard against a common fault. As the final logic connections are normally made in the safety circuits, it is likely that a number of critical cables, lines, etc., may be brought together. It is important that the possibility of short circuits or interaction faults between these critical points is minimized by adequate spacing and segregation. The rating of relay contacts, or their equivalent, is carefully considered to offset the danger of contacts welding or sticking due to overload currents.

In order to provide a required degree of reliability, the logic function may have to use more redundancy than the usual 2-out-of-3 operation allows. A typical application of the logic techniques employed on a power reactor is illustrated in Fig.6.

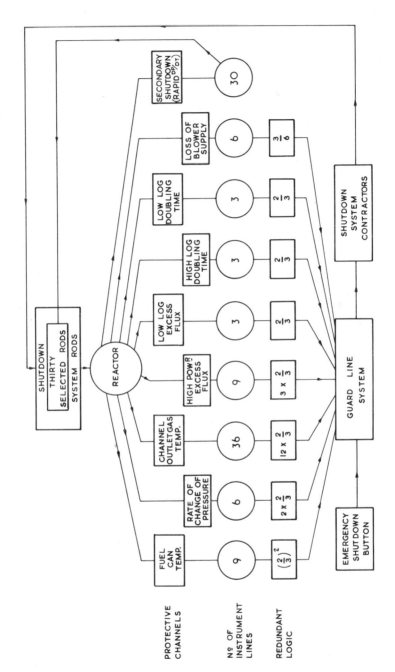

FIG.6 Schematic of typical automatic protective system for a power gas-cooled reactor.

5. Quantitative Approach to Control and Instrumentation Systems

Shutdown System

Normally, the protective requirement will have defined the degree of shutdown to be achieved in order to effect reactivity control. The method of shutdown may involve dumping the moderator, inserting absorbers, moving fuel, etc. Ideally, the shutdown system should be separate from the normal reactivity control function of the reactor so that the system does not have to be designed for a duplicate task with possible unsafe compromises ensuing. Generally speaking, this ideal is partly met with the reactor having only a limited number of actual safety shutdown mechanisms which are backed up following a trip action by the negative action of the normal reactivity control system. This is often known as the "primary system" since in some instances a separate diverse means of shutdown is used which is termed a "secondary system."

A power reactor is normally shut down by inserting a neutron absorbing material into the core. The insertion of the absorber is initiated by the trip signal from the safety circuits and provides sufficient quick-acting negative reactivity available at all times to give immediate return to the subcritical state. It is important that the amount of neutron absorber inserted into the reactor is sufficient to maintain a subcritical state, taking into account any changes to the system which might come about through the decay of fission products or through temperature changes in the fuel or coolant.

The minimum required negative insertion of reactivity and the reliability of the shutdown mechanisms define the amount of redundancy which is needed in the overall shutdown system. The system is normally divided, therefore, into a number of subsystems or individual elements. The total number is chosen to cater to the likely failure of individual subsystems and still leave an adequate margin in hand. Once again, as with any redundant system, care has to be taken to guard against the common fault. This aspect is particularly important where the insertion rate may be governed by some form of dynamic braking and a fault on the common bus-bar could give rise to the complete bank of control rods following a slower insertion rate for shutdown. The concept of each rod or absorber constituting a subcircuit with its own break contactor gives a high degree of assurance against the common-mode failure. Double contactor systems, which break the holding circuit in two places, may be employed to improve the reliability still further.

The neutron absorbers are normally introduced into the core under the force of gravity, but are sometimes assisted by additional hydraulic or mechanical devices.

Overall Performance

It is essential that the integrated system performance meets the original protective requirement under both normal and abnormal conditions. Once a reactor has been set up to operate as per design conditions, it is effectively protected on a "margin." The margin to trip is derived by various factors such as obtaining adequate response and accuracy by the protective system to limit overshoot and by having a sufficient size of margin to prevent spurious tripping due to normal operational procedure. Often, this concept of protecting on a margin is preserved by the use of excess margin to trip facilities which ensure that the trip settings bear a known and adequate relationship to the reactor operating conditions.

The various specifications for the system and components enable the failure rates of the equipment to be integrated to give the expected likelihood of failure of the system for each and every incident demand. The built-in redundancy combined with the specified periodicity of proof testing may then demonstrate that an adequate reliability can be achieved. In the ideal, it may be possible to give limits for the probability of the complete failure of the system such that it should not exceed a certain figure. Good design and specification of parts and inspection tend to eliminate as far as possible the type of fault arising from manufacture and assembly. Similarly, the wear-out type of fault is normally covered in the method of maintaining and testing the protective system. The random fault, however, may be the one on which the ultimate safety of the reactor depends, and it is important that adequate steps are taken to arrive at an indication of its likely occurrence rate.

Reliability Assessment of Protective Equipment

In carrying out a reliability analysis for an equipment, it is important to distinguish between different categories of failure; for example, safe equipment failures as opposed to dangerous equipment failures. Usually, safe equipment failures include those where an equipment fault causes its output to revert to the tripped condition, irrespective of input conditions. This aspect of performance is important from the plant availability point of view. Dangerous equipment failures are those where the output is maintained even when the input signal passes over to the "dangerous" side of the preset trip level. Such faults will affect the safety of the plant being controlled. These safe and dangerous faults can be either of

5. Quantitative Approach to Control and Instrumentation Systems

the revealed or unrevealed variety. The unrevealed dangerous faults category is usually the one of greatest concern since this can lie dormant on an equipment for a long period of time and would only be brought to light by a thorough proof checking procedure.

It is also expedient to assess the basic functioning of the equipment so as to ensure that it is capable of performing its intended task. A full reliability analysis then includes the assessment of the response time, accuracy, stability, etc., for every item of equipment. It is important to ensure that the components are operating within their specified maximum ratings. All these calculations involve the working out of voltages and currents in every electronic or electrical component, the signal magnitudes, stage gains, effects of component drifts with time, etc. This results in a detailed understanding of the performance characteristics for the equipment. Such a detailed understanding is an essential background to calculating the effects of component failure upon the equipment performance.

The effect of every relevant fault mode for each component is then assessed. Usually, an extreme fault analysis is made where the effects of open-circuit and short-circuit conditions on the electronic components are assessed. In practice, the magnitude of a fault for most components can be considered as infinitely variable between short circuit and open circuit. However, the extreme fault analysis will cover the effects of most of the intermediate faults sufficiently well to obtain valid results. The magnitude of an intermediate fault which will cause equipment failure will depend upon the function of the component in the equipment and a large range of these magnitudes covering many components will be applicable to an equipment of sufficient complexity. During the life of an equipment, intermediate as well as extreme component failure rates all cause equipment failure and they are all, therefore, recorded. Hence, by applying recorded failure rate figures to an extreme analysis, the performance of an equipment is usually predicted with sufficient accuracy for most safety evaluation purposes.

During the analysis, each component failure is placed into a fault category (i.e., safe or dangerous). Since there are many different possible fault categories, it is convenient to express them using a four-character code as follows:

First character
   S, fail-safe
   D, fail-dangerous
   C, calibration shift in the dangerous direction

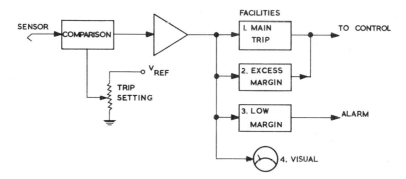

FIG.7 Example temperature trip amplifier.

Second character
This indicates which equipment facility is adversely affected by the fault, the facilities being numbered from 1, the most important being first.
Third character
u, a fault which remains unrevealed during normal operation
r, a fault which reveals itself
Fourth character
This indicates which facility reveals the fault, the numbering being identical to that used for the second character.

An example of the way in which numbers can be allocated to the equipment facilities is as follows:

(i)   Main trip
(ii) Excess-margin trip
(iii) Low-margin alarm
(iv) Indicating meter
(v)   Indicating lamps

In order to see how this coding system is used, an example of a temperature trip amplifier is taken, a block diagram of which is shown in Fig.7.

This trip amplifier consists of a comparator, an amplifier, a main trip function, an excess-margin trip function, a low-margin alarm, and visual indication. The input signal is compared with a trip setting and the difference signal is then amplified. If the signal is present, there is an output current to the control circuits which keeps the main trip relays energized. If the input becomes equal to the trip setting or higher than the trip setting, then the amplifier output is reduced to zero and the main trip function causes the control circuits to be deenergized. If the difference between the input and the trip setting is very large, then a large output signal from the amplifier will operate the excess-margin trip which will again pass a trip signal to the control circuits. Also, if the input approaches very closely to the trip setting, then the low-margin alarm will operate alarm circuits in the control room. Supposing something goes wrong with the trip setting such that its value is increased; this could be due to an open-circuit fault at the lower end of the trip setting potentiometer. This fault will automatically increase the difference between the input and the trip setting and will result in a large amplifier output. Also, in order for the main trip facility to trip the control circuits, the input signal will have to rise to a high level; this level will probably be higher than the maximum safe level for which the controlled plant is designed and, therefore, this is a dangerous fault (category D). The main trip facility of the trip amplifier is thus incapable of tripping the safety circuits at the correct input level; this is therefore a dangerous fault of the main trip facility (category D1). Without the excess-margin trip this would be an unrevealed fault condition except for the visual indication. In practice the meter is rarely inspected and this dangerous fault could, therefore, remain unrevealed on the equipment for some time. However, because of the presence of the excess-margin trip facility, the large difference signal will cause the excess-margin trip to operate. In this way, the dangerous fault condition becomes revealed by the excess-margin trip. This is therefore a D1r2 fault.

Another example would be an open-circuit fault at the top end of the potentiometer. In this case, the fault would be safe and would be immediately revealed by the operation of the main trip facility (S-r1 category).

Failure rate figures are then allocated for each component fault. When all the component faults have been assessed, the failure rates in each fault category are then added in order to arrive at a predicted failure rate for the equipment in each fault category. An example of the results of such an analysis is shown in Table IIA, with the totals shown in Table IIB. The actual effects of each of these equipment fault categories upon the plant being controlled will vary with each application, depending upon how the equipment is to be used. It is notable that the rate for the D1 faults is about 6.8% of the total failure rate. Also, the rate for the D1u category is about 3.2% of the total failure rate. These percentages are a measure of the success of the fail-safe design of the equipment.

In carrying out the theoretical prediction of the equipment failure rates in each fault category, it is usual to find

TABLE IIA

Failure Rates for Temperature Trip Amplifier

Fault category	Rate ($f/10^6$ hr)	Fault category	Rate ($f/10^6$ hr)
S-r1	53.89	D2r3	9.95
S-r2	2.1	D2r4	0.15
S-r3	7.4	D2u	6.0
S-r5	5.0	D2r2	0.35
S-u	15.84	D3u	3.3
D1r2	4.85	D4r4	5.8
D1r3	0.4	D5u	10.2
D1u	4.6	C1u	3.15
		C2u	6.79
		C3u	5.77

TABLE IIB

Total Failure Rates

	Rate	
Fault category	($f/10^6$ hr)	(f/yr)
All S	84.23	0.737
All D1	9.85	0.086
All D2 to D5	35.75	0.313
All C	15.71	0.138
Total for all faults	145.54	1.274

that a few of the faults are difficult to analyze. Such uncertainties can be due to preset potentiometer positions, the way in which a failed component can overload other components and cause further damage, the setting up and operating conditions existing at the time the fault occurs, or to the complexity of the theoretical fault analysis. Where the theoretical fault analysis is complex, the fault can be simulated practically on prototype equipment, the effects noted, and the correct category found. However, in cases where the effects depend upon the environmental and operational conditions, a practical simulation of the faults can be of dubious value, except for the particular equipment tested under the particular test conditions. Any uncertain fault which might cause fail-dangerous conditions should be placed in its worst probable fail-dangerous category in order to err on the safe side. The four-character code can be enclosed in brackets in these

5. Quantitative Approach to Control and Instrumentation Systems

TABLE III

Equipment Failure Rates

Equipment	Rate (f/yr)	
	Predicted	Observed
Temperature trip amplifier	2.8	2.3
Pulse rate measuring channel	17.8	17.4
Pneumatic pressure transmitter	0.45	0.76
Gamma monitor		
Dangerous failure	0.16	0.16
Spurious alarm	0.07	0.02
Other categories	0.15	0.16
Total	0.38	0.34

cases. By way of example, a transistorized equipment was assessed in which 629 different faults were analyzed. As a result of the theoretical prediction, only 15.3% of these faults required practical testing. The tests showed that about 3% of the predicted fault categories differed from the tested fault categories. Assuming that the faults which were not simulated were all correctly predicted, the theoretically predicted fault categories were correct for 97% of the faults studied.

Sometimes fail-dangerous conditions can result from two or more faults occurring together. By carrying out a detailed assessment such combinational faults can be highlighted. An example of a combinational fault is failure of the excess-margin trip facility (D2u fault) followed by failure of a main trip facility which would normally rely upon the excess-margin trip facility to reveal it (D1r2). The probabilities of such combinational faults can be calculated and if they prove to be significant compared with the probability of fail-dangerous faults due to single component failures, due allowance can be made in the final results of the assessment. Sometimes dangerous conditions can result from combinations of safe unrevealed or insignificant faults, and these are also usually highlighted by the detailed assessment.

Practical Failure Rates

Specific examples so far have dealt particularly with the assessment of electronic equipment. The methods discussed are equally applicable to equipments of the electrical, mechanical, pneumatic, hydraulic, and other types. The method in each case involves breaking down complex assemblies into easily recognizable parts for which data is either available or can

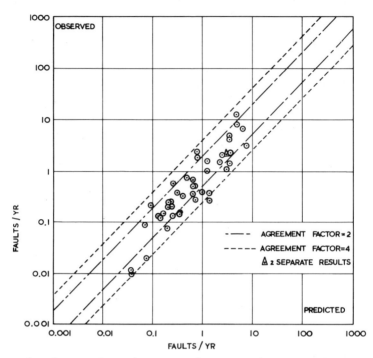

FIG.8 Observed against predicted equipment failure rates in 45 cases.

be obtained. Table III compares some practical failure rates for some equipments with the predicted failure rates.

The temperature trip amplifier used transistor circuits of comparatively early design. The pulse rate measuring channel is a more complex set of equipments using electronic valves resulting in a higher failure rate. The pneumatic pressure transmitter was made entirely of pneumatic and mechanical components. In this case the observed failure rate is rather higher than the predicted value, due mainly to a rather high blockage rate in the restrictors. This was caused by oil and dust contamination in the compressed air supply. The failure rates for the gamma monitor are divided into failure rates relating to a few basic fault categories. The largest difference occurred in connection with the spurious alarm rate. The ratio of predicted to observed failure rate is 3 1/2:1 in this case.

Many other equipments have had their failure rates predicted and in some cases observed failure rates are available. Figure 8 shows the relationships between predicted and observed failure rates in 45 different cases. The failure rates at the upper end of the scale are mainly electronic valve equipments

5. Quantitative Approach to Control and Instrumentation Systems

which have shown failure rates in the region of 1 to 10 f/yr. Transistor equipments generally lie in the region of 0.1 to 1 f/yr. It is expected that transistor equipments of more modern design using high-quality components and careful manufacturing techniques may lie in the region of 0.01 to 0.1 f/yr. Pneumatic and mechanical instruments have been assessed and their failure rates have usually been in the region of 0.1 to 1 f/yr; these instruments are generally of simple design, where the number of components used is relatively small. The points shown between 0.01 and 0.1 f/yr in Fig.8 represent failure rates in specific fault categories for particular instruments: typically the spurious alarm rate for nuclear radiation warning instruments. The scatter in the results of Fig.8 are due to the following.

(i) Errors in the prediction method: the assumed failure rates for the components are based upon previous experience with other equipment and will not apply accurately to the equipment being assessed, etc.

(ii) Errors in the observed values: these errors occur due to difficulty in precisely determining which component failed and to the possibility that the precise number of failures may not be accurately reported, etc.

In order to see whether there was an easily recognized statistical distribution in these errors, a plot was made of the observed proportionate frequency of the ratio "observed failure rate to predicted failure rate" less than a factor R. This plot is shown in Fig.9 on a probability scale and using a logarithmic scale for R. The curve indicates the following.

(i) The ratio of observed to predicted failure rate appears to follow a log/normal distribution.

(ii) The median value for R (i.e., P = 50%) is 0.76 which indicates that, on average, the predictions are pessimistic by about 30%.

(iii) The chance of the ratio being within a factor of 2 of 0.76 is 70%.

(iv) The chance of the ratio being within a factor of 4 of 0.76 is 96%.

Data Collection

As components and equipments become more reliable with improvement in manufacturing techniques and the establishment of better quality control, the need for collecting reliability data will continue. Reliability data is becoming increasingly important as plants become more complex and costly. A number of failure rate data schemes are already functioning in the

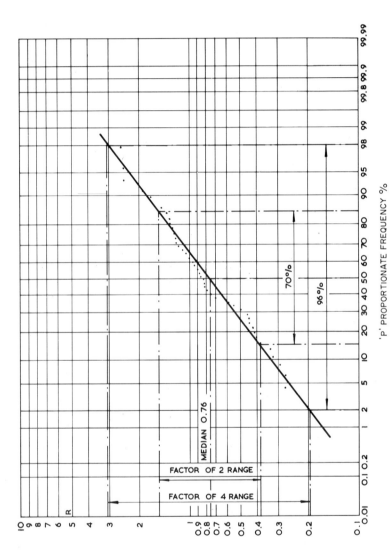

FIG.9 Observed proportionate frequency of the ratio, observed failure rate/predicted failure rate, less than R.

5. Quantitative Approach to Control and Instrumentation Systems 95

military field and in the more important sections of industry. It is convenient if the data can be accurately collected in conjunction with the normal operational and maintenance methods which are necessary for the running of the plant. In some cases it may be possible to use the same maintenance/job cards, possibly with slight modification. A suggested method of collecting reliability data can be found in Eames [4], and a brief description of a data bank system is given in Ablitt [5]. Reliability data collection (following prediction methods) is an essential part of a reliability analysis. This arises because of the following.

(i) Where a high degree of safety is required, the practical figures are required as a confirmation of the original predictions. The data only becomes complete at the end of the working life of the equipment.

(ii) The failure rate data must be continually updated so that it can be applied to new equipment and system designs.

Prediction techniques must be applied with care. For example, some of the assumptions which may have to be taken into account in the reliability analyses discussed are: (i) failures are random; (ii) failures are independent; (iii) no compensating failures; (iv) testing time is negligible; and (v) repair of equipments is perfect.

Failures are random. This implies that whatever the cause of failure the actual failures appear on mass to follow the laws of chance and are capable of being represented by a statistical distribution. When the exponential distribution is used, it is assumed that the equipments are operating in the useful life phase where the average failure rate is constant over a large number of failures.

Failures are independent. This implies that failures cannot propagate from component to component. A failure of a component affects only that particular component and nothing else.

No compensating failures. This is the assumption that two wrongs do not make a right, i.e., that two failures cannot cancel each other out so as to present an apparent picture of the equipment working normally.

Testing time is negligible. Generally, while an equipment is being tested it cannot be functioning normally. If the testing is very frequent, then the equipment may be nearly always on test and rarely fulfilling its intended purpose. It is assumed that such equipment "dead time" is negligible.

Repair of equipments is perfect. This assumes that whenever an equipment is inspected or tested, it is restored to perfect working condition. Every fault or defect is discovered and the necessary repair carried out to restore the equipment to "mint" condition.

Only if all these assumptions, and many others like them, can be established as having a probability of failure which is orders smaller than the numbers revealed by the simple reliability calculations, can the calculations be considered realistic. For probability numbers down to the order of 10^{-2} it is generally fairly easy to substantiate the assumptions for single equipments. Below these values it usually becomes necessary to refine the reliability calculations to take other contributing factors into account.

With appropriate care, the agreement which can usually be obtained between predicted and observed results is sufficient for most reliability evaluation purposes. The methods of analysis outlined are proving increasingly useful in the reliability analysis of equipments in many fields of application.

Failure Probabilities of Components and Systems

In the foregoing, components of an automatic protective system have been described and the importance of different modes of failure has been discussed. For some purposes, e.g., in some decisions on maintenance strategy, the failure rate parameter may be sufficient, as the need for maintenance and/or repair will obviously depend on the number and types of failures occurring in a given period of time. It is necessary, however, once the inherent capability of the system has been established, to keep as the overall aim the determination of the degree of success (reliability) the system will have in reducing the reactor to a safe state should abnormal conditions arise.

Since the pattern of failures at system level may be a complex function of time or environment, particularly when the design incorporates redundancy and diversity to ensure that it can still function even though some components have failed, it is necessary to think in terms of overall "probability of failure" rather than the "mean failure rate" of the system. Further, to take account of unrevealed dangerous failures it may be more important to estimate the amount of time or proportion of time for which the system is incapable of functioning, i.e., its "fractional dead time." System failures are generally compounded from a pattern of random events which can be represented by a statistical probability density function of the following form

$$f(t) = \theta(t) \exp[-\int \theta(t)\, dt] \tag{2}$$

where $f(t)$ is the probability density function (pdf) for the distribution of the random events in time, and $\theta(t)$ the event-rate function. Only when $\theta(t)$ is constant over the time or under the conditions of interest can the system failure be interpreted as having a "mean rate" of occurrence. If $\theta(t)$ has a constant value θ, then the general pdf reduces to the familiar exponential form:

$$f(t) = \theta e^{-\theta t} \qquad (3)$$

Another reliability parameter of general interest is the "probability of failure" such as the probability that a protective system has failed by the end of a certain length of time. This may be given by the cumulative probability function which is obtained by integrating the pdf over the time range of interest, for instance,

$$p(t) = \int_0^t f(t) \, dt \qquad (4)$$

which, for the particular case of the exponential pdf given in Eq.(3), reduces to

$$p(t) = 1 - e^{-\theta t} \qquad (5)$$

and this will be recognized as the form of probability function used earlier [see Eq.(1), page 74]. Fractional dead time can be calculated from this basic relationship as described in Refs. [1] and [3]. The importance of system fractional dead time lies in the fact that an automatic protective system installed in a nuclear reactor normally plays a quiescent role. It is only when a demand arises, perhaps due to a reactor fault condition, loss of electrical supplies, or some other predictable or unpredictable cause, that it is required to operate. Such demands are generally looked upon as having randomly distributed times of occurrence. The automatic protective system will fail if, when the demand occurs, it contains some unrevealed system fault or faults which prevent correct operation. Unrevealed faults are found by periodic proof testing and corrected immediately to minimize dead time. Fractional dead time provides an index of the probability that the system will be found in the failed state.

It is unlikely that failure characteristics of a complete automatic protective system will be fully known and capable of being expressed in probabilistic terms. The more likely situation is that only the failure characteristics of the various individual items or component parts of which the complete system is comprised are known. These individual parts are functionally combined in what is often a complex arrangement in order to produce the complete functioning of the system. There is a need, therefore, for a method of synthesis which enables the individual failure characteristics to be combined in a similar functional manner in order to arrive at the failure characteristic for the overall protective system. Such a requirement for a method of synthesis is similar to

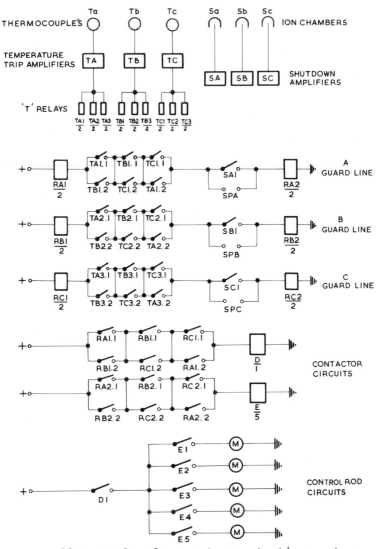

FIG.10 Example of a reactor protective system.

that described earlier in connection with an item of equipment and its component parts.

In synthesizing a complete system account has to be taken of the types of failure and the effects they cause. Failures may be partial, that is, a device may operate but in a manner which is outside the normal specification limits. Failures may be catastrophic and result in complete loss of operation or performance. Failures may affect individual components,

5. Quantitative Approach to Control and Instrumentation Systems

complete devices, or entire subsystems. The pattern and magnitude of failures may vary with time, space, and environment. Repairs will depend upon whether the failures are immediately revealed or remain unrevealed for a period. In addition, restorations may depend upon the availability of labor, stock levels, and spare parts. Each type of failure and its subsequent repair, if any, will generally lead to a different chain of events and a different overall result. In order to establish the appropriate probability expressions for each chain of events, it is useful to convert the protective system functional diagram into a logic sequence diagram.

By way of illustration, Fig.10 shows a simplified version of the functional diagram for a reactor automatic protective system and Fig.11 shows one form of the corresponding logic sequence diagram. In Fig.11, the smaller circles represent the reliability characteristics of each device in the system, the larger circles containing the annotations such as 2/3 represent the logical functions, and the connecting lines represent the flow of required information. The annotation 2/3 means that any 2 or more of the 3 inputs require to be present before an output is produced.

It is now necessary to combine the reliability characteristics of each device according to the logical arrangement shown in the appropriate logic sequence diagram. There are a number of ways in which this may be done. For very simple systems which consist of only a few functional blocks, the appropriate probability combinations may be obtained by constructing the relevant "truth table." The probabilities may also be combined algebraically using the relationships developed [1], and for complex systems a computer program could be used [6].

With a given failure rate for equipment, the probability of failing dangerously can be reduced to some extent by shortening the proof-test interval. It is desirable that each complete channel of protection is proof tested by primary injection methods as far as is practicable. The use of the reactor as a source of primary injection normally gives a more direct and overall test than signals injected into part of the equipment. Each channel of protection, in its entirety, normally has provision for proof testing, not only to show its operation but also its time of response.

It is important that new and novel designs be subjected to extensive environmental proof testing under controlled conditions. It is equally important that sufficient sample testing be carried out under actual operating conditions to substantiate any improvement in the reliability claimed.

Generally proof testing is closely associated with maintenance, and this may be defined as the art of ensuring that

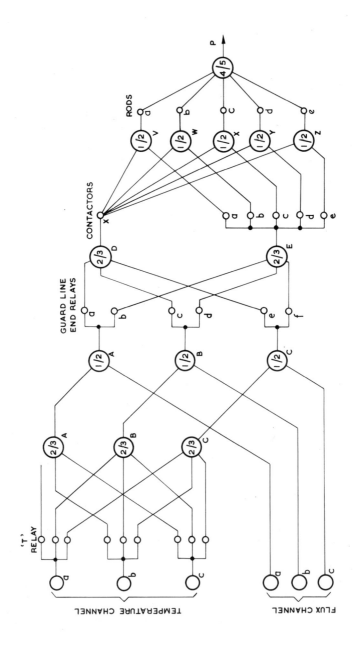

FIG.11 Logic sequence diagram for reactor protective system.

5. Quantitative Approach to Control and Instrumentation Systems

the system keeps within a set of predetermined limits. In general, maintenance is based on principles employing preventive techniques, the frequency being governed by past experience and the feedback of fault statistics together with comprehensive records.

Analysis of Shutdown System

Similar techniques have been applied in predicting failure rates and fractional dead times for shutdown systems employing mechanical items such as valves, liquid absorbers, and pneumatics as well as for the more conventional electromagnetic clutch systems. Usually for such systems a boundary approach is most appropriate because they form part of the reactor core design and employ high redundancy for operational as well as safety reasons.

Analysis of experience to date indicates that a failure probability per demand of somewhere between 10^{-2} and 10^{-3} per element is easily achieved for the electromagnetic clutch type and a boundary figure for assessment purposes of 10^{-2} per element is not unreasonable. Care must be taken, however, to ensure that common fault modes which might be caused by wear-out, distortion due to irradiation and thermal effects, core shift, etc., are minimal and preferably can be shown to be of a lower order of probability than the system requirement. Because of the difficulty of demonstrating such low probabilities, the assessor is continually seeking diversity for shutdown by provision of alternative or secondary shutdown systems.

In view of its novel features and importance for fast shutdown, a liquid shutdown system (Fig.12) was analyzed in detail using predictive techniques. A summary of the main features yielded the following fail-to-danger figures on a single shutdown loop basis. All the elements in one loop can be considered in series since failure of one element would imply failure of the loop.

Element	Fractional dead time (based on 12-week proof test interval)
Monitoring system	1.25×10^{-5}
Pipework system	3.5×10^{-3}
Main shutdown valve	1.2×10^{-2}
Total	1.55×10^{-2}

Information from test results on prototype and production equipment was analyzed with the following results:

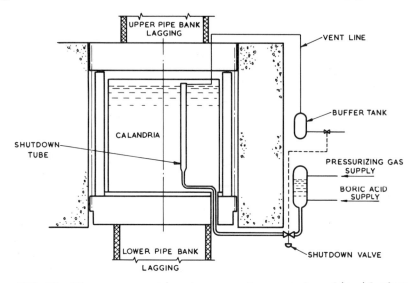

FIG.12 Steam generating heavy water reactor liquid shutdown system (single loop).

Element	Chance of failure on demand
Complete system (excluding valves)	4×10^{-3}
Auxiliary valve	3×10^{-3}
Main shutdown valve	1×10^{-2}
Total	1.7×10^{-2}

These results compare favorably with prediction. Such figures enabled an estimate to be made of the chance of system failure on demand based on the fractional dead time for each loop.

Loops required for success	Chance of system failure on demand
12-out-of-12	1.7×10^{-1}
11-out-of-12	4.5×10^{-3}
10-out-of-12	1.7×10^{-4}
9-out-of-12	4.6×10^{-6}

In considering these figures, it should be appreciated that the ways in which the system is operated, maintained, and tested may be the factors of prime importance in achieving the standards required. In making such estimates, close attention has to be given to all relevant administrative procedures to ensure that possible common fault modes, e.g., in filling with the wrong strength of the poison absorber, are eliminated as far as possible.

5. Quantitative Approach to Control and Instrumentation Systems 103

Failures recorded to date show that overall failure rates are within a factor of two lower than predicted. During commissioning a common fault mode was discovered in that the material for the seating of the main valves was found to be unsuitable for the environment, which was at a higher temperature than originally estimated, and there was an occurrence of the poison being diluted. Such failures were not catastrophic and could clearly be classed as partial failures. It is noteworthy that these and other wear-out failures were disclosed by routine examination, indicating that administrative procedures are being applied and are being shown to be adequate in averting what could be dangerous failures if allowed to continue uncorrected.

Role of the Operator and Computer

Human factors have to be given consideration in assessing the reliability of any technological system. Human influence is felt in design, commissioning, routine operation, and maintenance. At the design stage due place has to be given to ergonomics [7] within the context of the whole system so that man/machine interaction is as efficient as possible. Although automation is being extensively applied, there arise situations where the operator is employed essentially as a high-level information processor and decision maker.

On the other hand, it is quite common to find that "human maloperation" is quoted as a frequent cause of faults in reactor systems. Automatic protective systems, which are generally required to be fast acting, i.e., have a response time requirement shorter than can be reasonably met by an operator pressing a button, are designed in such a fashion as to minimize maloperation by use of redundancy and diversity, restricted access, etc. Strict operating rules aimed at minimizing interference which would reduce safety and reliability are a necessary part of reactor operation.

Table IV gives a general comparison between the suitability of man and machine for various functions. As would be expected, machines are best for performing any activities which can be specified precisely because they are superior in terms of speed and reliability. Man, however, excels in tasks involving inference and extrapolation and in some cases decision making under uncertainty. It should always be remembered that so far as the core of a nuclear reactor is concerned, the operator has to rely on indications provided by the installed instruments and response time to these could be measured in minutes before an appropriate action is performed. The trend in providing information to the operator is more and more towards centralized control and aids are provided for processing data

TABLE IV

Relative Advantages of Men and Machines[a]

Task parameter	Machine	Man
Speed	Much superior	Lag one sec
Power	Consistent at any level Large, constant standard forces	2 hp for about 10 sec 0.5 hp for a few minutes 0.2 hp for continuous work over a day
Consistency	Ideal for routine, repetition, precision	Not reliable, should be monitored by machine
Complex activities	Multichannel	Single channel
Memory	Best for literal reproduction and short-term storage	Large-store multiple access. Better for principles and strategies
Reasoning	Good deductive	Good inductive
Computation	Fast, accurate Poor at error correction	Slow. Subject to error Good at error correction
Input sensitivity	Some outside-human senses, e.g., radio-activity	Wide range (10^{12}) and variety of stimuli dealt with by one unit, e.g., eye deals with relative location movement and color
	Insensitive to extraneous stimuli Poor for pattern detection	Affected by heat, cold, noise, and vibration Good at pattern detection. Can detect signals in high noise levels
Overload reliability	Sudden breakdown	Graceful degradation
Intelligence	None	Can deal with unpredicted and unpredictable. Can anticipate
Manipulative abilities	Specific	Great versatility

[a] After Singleton [7].

5. Quantitative Approach to Control and Instrumentation Systems 105

before presentation. Frequently, such processing is based on a digital computer which affords facilities for a broadband sampling display or a more detailed one, e.g., a display of temperature profiles across the core. Attention is drawn to abnormal conditions by special alarm displays which can give broad indications of urgency and source of fault. Displays which come on cathode ray tubes can be qualitative, quantitative, or representational (e.g., mimic diagram). The computer provides a centralized store of information on plant status and availability. It can also be used for calculating direct from the data provided on-line and also use back history information in fuel burn-up calculations. The computer is also being used to replace the operator in automatic start-up.

It is difficult to assess the contributions of the computer and operator [8] in terms of quantitative reliability. The first is subject to "software" errors connected with programming, and it is mainly the software problem which needs resolution before computers might be used in an automatic protective role. Human reliability is dependent on too many factors, e.g., state of mind, motivation, social aspects, and various forms of stress. Most of the information available on human tasks is often abstract and simplified, and the general environment is atypical. Data on human performance in real situations might be obtainable by examining the back history information provided by the computer monitoring the state of the plant.

General Methods of Assessment
of Overall Performance

The steps in a safety and reliability assessment are summarized in Fig.13, which is in essence a form of critical thinking. This starts with a statement of the problem and proceeds through the substantiation of each stage to the solution which is then compared with the statement first given. Throughout the assessment the following questions will recur, and the assessor will be satisfying himself as to the validity of the answers.

(a) Is the quantity the correct one to be measured and, bearing in mind all the assumptions, are we measuring what we think we are?

(b) Is the quantity to be measured a time variant or is it constant during the time required for measurement?

(c) What accuracy can be guaranteed from this measurement?

(d) What is the time delay in measuring and having the

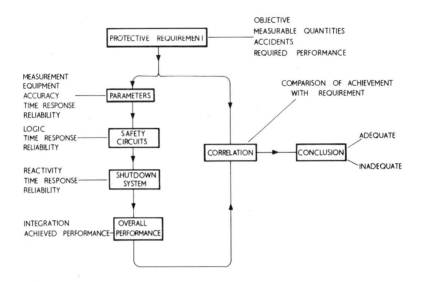

FIG.13 Steps in safety assessment.

information available at the output?
(e) How reliable is the equipment provided?

The first step in the assessment is to combine the detailed appraisals and to assess the overall performance of the automatic protective system with direct reference to the possible reactor faults against which protection is claimed. The assessor therefore (i) summarizes the integrated performance of each protective channel with reference to accuracy, response, and reliability, and (ii) correlates the relevant integrated performance with the appropriate reactor fault analyses which will have been considered in the protective requirement.

It should be noted that the detailed performance of the various components when viewed as a system must be considered under both normal and abnormal conditions. In addition, the system must be investigated to show that it has the capability of functioning correctly and the extent to which this may be demonstrated. Where information is not given on various factors or where the assessor is unable to accept the information that has been given, the boundary approach is very useful. In this approach the assessor may use his critical judgment to estimate boundary performance values in order to complete the exercise. It is essential that any such information should be clearly identified as being the work of the assessor.

References

1. Green, A. E., and Bourne, A. J. "Reliability Technology." Wiley, New York, 1972.
2. Green, A. E., and Bourne, A. J. Reliability considerations for automatic protective systems, *Nucl. Eng. 10*, No. 111 (August 1965).
3. Bourne, A. J. A criterion for the reliability assessment of protective systems, *Control 11*, No. 112 (October 1967).
4. Eames, A. R. Data Store Requirements Arising Out of Reliability Analyses, UKAEA Rep. No. AHSB(S)R138 (1967).
5. Ablitt, J. F. An Introduction to the "SYREL" Reliability Data Bank, UKAEA Rep. No. SRS/GR/14 (1973).
6. Woodcock, E. R. The Calculation of the Reliability of Systems—The Program NOTED, UKAEA Rep. No. AHSB(S)R153 (1968).
7. Singleton, W. T., Easterby, R. S., and Whitfield, D. (eds.). "Proceedings of Conference on the Human Operator in Complex Systems," Taylor and Francis, London, 1967.
8. Swain, A. D. Human Reliability Assessment in Nuclear Reactor Plants, Sandia Laboratories, New Mexico, Monograph SC-R-69-1236 (1969).

6
The Reliability of Heat Removal Systems
F. M. Davies

Introduction

Reliability has always been recognized by engineers as a necessary feature of good design. In the past it has been achieved by applying experience with sound judgment, but the design process has been subjective and qualitative.

The advent of advanced technologies with associated high risks has now given considerable impetus to the adoption of quantitative reliability analysis techniques. The space and atomic energy industries are prime examples in providing this impetus. Furthermore, once risk targets had been defined numerical analysis methods logically followed. In the UKAEA nuclear industry the Farmer criteria [1] defined objectives which set the scene for considerable activity in the formulation and adoption of reliability analysis techniques.

This has meant the addition of a further dimension in the design process. Hitherto designers had confined themselves to their traditional activities in assessing plant performance, in arriving at a design compromise when evaluating various choices available to them. Now the design criteria have been expanded so that designers ask themselves additional questions such as: What is the chance of failure of a particular engineered safeguard? Will it meet the nominated risk targets? Current techniques allow an improvement in the precision with which answers to such questions may be made, but at the same time it must be admitted that with the present state of the art one may only be able to predict answers within an order

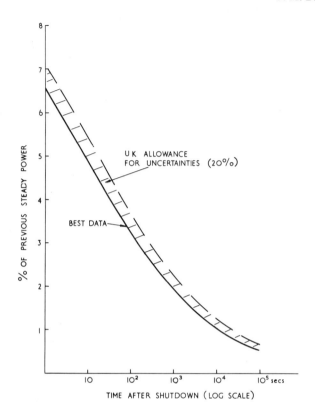

FIG.1 Variation of afterheat with time. Data include fission fragments only, and not plutonium precursors.

of magnitude. It is anticipated, however, that improvements in techniques and data will in the future lead to greater accuracy in design predictions.

Heat Sources

As a heat source the nuclear reactor differs from most other power systems because it is not possible to turn off the heat source completely. The afterheat following a reactor trip arises from the radioactive decay of fission products produced during power operation. Figure 1 illustrates the best estimates for total decay heat versus decay time expressed as a percentage of full power assuming infinite reactor irradiation.

Hence, the total decay heat is 6.3% of full power at shutdown, reducing to half this value in 95 sec, and thereafter

6. The Reliability of Heat Removal Systems

decreasing to 1% in 2.5 hr. At 24 hr after shutdown the decay heat has reduced to about 0.5% of full power. With current reactors the heat produced can be considerable. For example, if the heat rating is 2500 MW, the decay heat is still 12.5 MW after one day.

The deposition of the decay heat will depend upon the particular reactor under consideration. However, all the beta heating and about 80% of the gamma heating will be deposited in the fuel, the remainder of the gamma heating being deposited in the core structure and outside the core. Hence, it may be assumed that over 90% of the heat goes into the fuel, but for all practical purposes calculations usually assume 100%. Apart from the decay heat during the initial temperature transients, other sources of heat may be realized under accident conditions [29], such as heat arising from metal/water reactions and the decay of actinides.

Failure to remove any of the decay heat following reactor trip would inevitably result in fuel failure and the release of some activity from the reactor to the core. The fraction of total activity released to the environment would depend upon the type of accident situation under study, the degree of containment afforded, and the mechanisms involved in the activity dispersion process. All these factors combined with the frequency of accident situations may be embodied in a risk criterion as discussed in Chapter 4. A typical criterion is illustrated in Fig.1 (Chapter 4), where the acceptable frequency for various activity releases (Ci ^{131}I) is defined by a simple curve. Whereas this particular criterion relates to the ^{131}I activity, similar criteria may be evolved for other types of activity release.

Reliability Targets

In setting reliability targets it is first of all necessary to postulate hypothetical accident conditions and assess their consequences. The acceptable frequency for a particular accident may then be determined from the risk criterion. For example, if the assessment procedure showed that 10^4 Ci ^{131}I would be released, then the acceptable accident frequency may be defined as 10^{-4}/yr. In the case of this particular accident, therefore, the system failure probability would be defined by the equation

Acceptable accident frequency = System demand frequency × System failure probability

If in this example the demand frequency was 10 demands per year, then the system failure probability target for this particular accident would be 10^{-5}.

There is a very broad frequency spectrum associated with the initiating events. Following any scheduled or spurious reactor trip, a demand is put on the shutdown cooling system, and in the early life of the plant this could amount to some 20 demands per annum. On the other hand, fracture of a high-pressure coolant circuit might carry a risk of 10^{-3}-10^{-4}/yr. This means that the designer's efforts should be concentrated on producing a high standard of reliability for heat removal following shutdown under normal conditions. The actual target level, whether it is 10^{-4} or 10^{-6} or 10^{-8}, will depend upon the analysis of particular reactor systems and will depend upon the safeguards provided to limit activity release. It also means that for highly improbable accidents, such as loss of coolant accidents in pressurized systems, the cooling performance requirements need not be so critical. The limiting case of pressure vessel failure is the ultimate accident where coolant cannot be contained and cooling systems are rendered ineffectual.

There are also good economic reasons for requiring highly reliable systems for normal operation. Clearly, however, there is a limit to the reliability target based on both economic and safety requirements and indeed what is practicably achievable. The current view within the industry would seem to suggest a target failure probability for well-designed and maintained systems in the 10^{-5}-10^{-7} range.

One of the prime reasons for this view is the presence of common fault modes. Theoretically speaking, very highly reliable systems could be provided by redundant systems, but there is a limit to the claims that can be made because ultimately systems rely on common services such as water or electricity supplies.

Reliability Evaluation

There is a wealth of literature currently available on reliability and its application to the evaluations of practical systems. A short reading list is given in Refs. [2-18] to which the reader may refer.

In this monograph, reliability is defined in Chapter 5 as the "probability that the item will perform in the manner desired for a specified period of time under the specified environmental condition." Hence, before reliability analysis can proceed, plant performance must be defined. In the context of emergency cooling systems, this means the calculation of the minimum cooling requirements in relation to system component temperatures beyond which it is unsafe to proceed. This calculation will also be required to show the grace period

6. The Reliability of Heat Removal Systems 113

allowable before switching in the emergency cooling system and
the time interval over which cooling is required.
Various methods for reliability analysis have been tried;
namely, truth tables, fault trees, logic diagrams, and network
synthesis. Manipulation of the techniques may be either manual
or by means of computer programs [19-24]. The analysis is
straightforward and embodies the following major steps.

(a) A definition of the problem together with assumptions,
failure criteria, and system diagrams.
(b) Setting up network logic diagrams or fault trees.
(c) Assimilation of input data such as fault rates, inspection intervals, and repair times.
(d) Computation of failure probabilities.

A major fraction of the work is devoted to the preparation
of the logic diagrams. Broadly speaking, the preparation of
such diagrams fall into two distinct methods. The fault tree
method is essentially a definition of system failure modes,
whereas network synthesis involves setting up a logic diagram
of how the plant under study normally works, i.e., a success
diagram. These procedures are discussed in Ref. [7].
Hand calculations may be made for systems with a small
number of components or for dealing with subsystems, but where
a large number of components is involved, then computer solutions are sought. It will be appreciated that the truth table
method becomes unwieldy as the number of components increases.
If, for example, there are 100 components, then 2^{100} plant
states have to be considered. Even with computers, programs
have to be devised to reject the low probability states to
reduce the problem to manageable proportions.
The PREP and KITT computer programs [23,21] have been
used extensively to evaluate the reliability characteristics
of engineered safety features of American nuclear reactors.
The PREP computer program is designed to accept a description
of the system fault tree in a format which is natural for the
engineer familiar with standard fault tree terminology, to
generate a logical equivalent of that fault tree, and then to
obtain the minimal cut sets of the system. These minimal cut
sets can then be analyzed by one of the KITT programs to obtain the point reliability information about the system. The
propagation of errors from component to system level in the
fault tree can be calculated by the SAMPLE computer program
which uses the Monte Carlo sampling technique to generate the
error distribution of the probability of the top event occurring.
In the UK the NOTED program [19] has been used for similar
reliability evaluations. It requires the preparation of a
logic network consisting of a number of switches connected by

links along which a signal can pass. Signals may pass freely along these links but can be impeded at the switches—so called by analogy with electric circuits. Associated with each switch, except those labeled START and FINISH, is the probability that a signal fails to leave the switch, given that a signal enters it. Three types of logic gates are handled; namely, AND, OR, and mutually exclusive inputs.

Time-dependent probabilities may be simulated by various law types, some of which are discussed in the Appendix. The probability of failure due to revealed and unrevealed faults may be treated in addition to the probability of repair. Time delays in operation may also be incorporated.

System Testing to Substantiate Predictions

Testing to substantiate reliability predictions is viewed as an overall commitment starting with the design process and continuing right through the manufacturing stage, plant commissioning, early operational phases, and ultimately the service life of the plant. The reasons for testing, therefore, stem from the need to confirm: (a) quality control and inspections of plant components; (b) that the plant performance is within the design specification; (c) that the plant as constructed will function as designed; and (d) that the initial reliability predictions can be realized with a certain degree of confidence which is at the worst maintained and at the best increased throughout the service life of the plant.

The substantiation of plant reliability is a growth process which should start from tests on components leaving the production line, proceed through organized tests on the system during commissioning and early reactor operation, and continue throughout the reactor life from the accurate and systematic recording of fault data. In this way, the reliability estimates start with tentative values at low confidence and become firmer at higher confidence levels as the system proceeds through its life. Thus, practical and theoretical approaches are used to maximum advantage during all phases by cross-fertilization of the results of the two approaches.

Hence, the general approach to the substantiation of reliability is as follows.

(a) From the theoretical analysis concentrate on critical components which make the most contribution to system unreliability, by virtue of high fault rates or high repair times.

(b) Complete shakedown tests to remove rogue components.

(c) Carry out a limited series of reliability trials to confirm predictions, albeit at a reduced confidence level.

(d) Continue reliability trials during the life of the

6. The Reliability of Heat Removal Systems

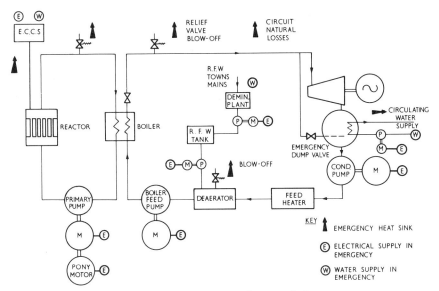

FIG.2 Emergency heat sinks.

plant to improve confidence in the predictions.

(e) Continue surveillance of plant component failure rate/repair time data.

(f) Carry out continuous evaluation of data to confirm predictions of random failure probability and detection of common mode faults which could invalidate predictions.

Basic Methods for Heat Rejection

A simplified diagram of the main and emergency heat sinks in a nuclear reactor system is shown in Fig.2. The reactor primary coolant circuit usually comprises a number of closed loops through which coolant is pumped to reject heat via the heat exchangers to a common steam system. The steam system is also shown as a closed loop where the major part of the energy transfer is to the circulating water supply in cooling the condensate so that it can be pumped back into the system. Electrical power is shown delivered to the main electrical system from a turbo alternator.

Since turbo alternators are unsuitable for use below 50% nominal rating, decay heat must be rejected through one or more of the following paths.

(a) by natural heat losses through the primary tank or at any point in the system;

(b) by means of an auxiliary emergency core cooling system;

(c) by steam dump; and
(d) by blowing off steam.

All these features are present to varying degrees on current reactor systems. For the Pickering reactor [25], it is said that even in the event of failure of emergency core cooling, core meltdown does not result because the relatively cool moderator surrounding the pressure tubes acts as a heat sink. Phenix, the French fast reactor, employs water cooling external to the reactor guard vessel and air cooling arrangements for the steam generators [26]. The British prototype fast reactor relies on an emergency steam dump system combined with a highly reliable decay heat rejection system using dip coils in the primary sodium pool. Special arrangements are provided for the steam-generating heavy-water reactor to provide core cooling following a rupture at any point in the pressure circuit. A high-pressure storage tank delivers cooling water to each fuel cluster via a central sparge pipe in the depressurizing phase followed by long-term cooling water supplied by gravity feed from the demineralized water storage tank which can be kept topped up from one of several sources. Under these accident conditions heat may ultimately be rejected to the suppression pond which itself requires cooling. Under normal operational shutdown conditions the same pond is used for removing decay heat. Main features of the system are shown in Fig. 3.

The features illustrated in Fig. 2, however, point to the directions in which engineering solutions lie, and some of the key problems which may be encountered in arriving at those solutions. The emergency heat removal problem may be divided into two distinct parts, namely, heat rejection from the core through the primary circuit(s) to the boiler(s), and then from the boiler(s) to the environment.

For scheduled shutdowns reactor power can be reduced in a controlled manner to maintain system temperatures sensibly constant and then reduce them gradually to prevent thermal shocks. Where a reactor trip occurs due to a fault arising in the system, then a contoured trip may be used in an attempt to achieve the same desirable result. In some circumstances, however, such as the case where the primary pumps run down following a trip due to main supply failure, the fuel/clad temperature would rise both on account of the stored energy in the fuel and the adjustment of temperature gradients consistent with the reactor now operating at decay heat level. The rate of rise of clad temperature may, however, be modified by fitting the pumps with flywheels to prolong the shutdown period. The key question which arises, however, is whether forced convection is necessary following primary pump rundown

6. The Reliability of Heat Removal Systems

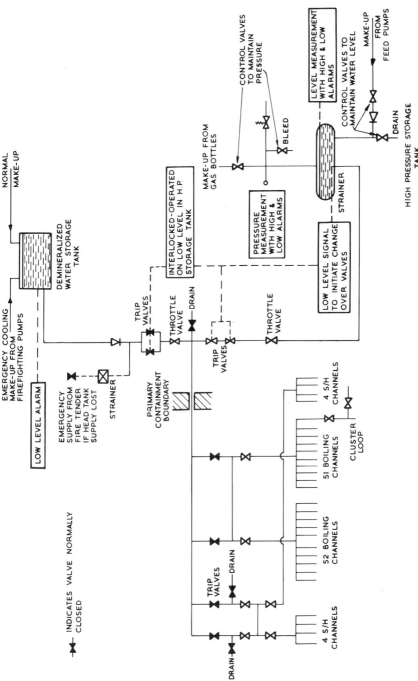

FIG.3 Flow diagram for channel emergency cooling system. No connections are shown for requirements other than emergency cooling (e.g., start-up, drum inspection, or emergency feed).

or whether one can rely on natural circulation. Natural circulation can be promoted by siting the boilers above the reactor core so that the density differences between the hot and cold legs can overcome the circuit frictional losses under flow conditions. Some systems may possess a peculiarity which precludes this. The PFR exhibits some difficulties in this direction due to the particular arrangement of the primary circuit as this is a pool type reactor. For this reason pony motor auxiliary drives are employed to prevent flow stagnation in the core should the main pumps stop.

In other reactor systems pony motors are deemed necessary to provide forced convection in the long term because of doubts concerning the integrity of primary pump(s) drives, particularly where they are steam driven. Overall, it seems that for one reason or another designers deem it necessary to fit auxiliary drives to primary pump(s) and the reliability of such drives has to be guaranteed to a minimum standard. Invariably for this and other reasons nuclear stations incorporate a so-called guaranteed auxiliary electrical supply relying on diesel generators, the reliability of which has to be assessed. Ultimately, high-grade reactor heat has to be rejected as low-grade heat to the environment, and for this reason copious water supplies are required, failure of which represents a common mode fault.

In relation to steam systems because of the large amount of stored energy available following off loading of the turbine, it is attractive to consider whether the reliability of the steam system can be brought up to the high reliability standards required to energize the pump drives. Such schemes [28] are currently under active consideration, particularly in relation to the question of providing continuity of cooling following a reactor trip. In general, however, it may be said that without special attention to engineering detail the standards attained by commercial steam systems fall short of those required for decay heat removal in nuclear power systems.

The diversity of the various approaches to fast reactor decay heat removal is evident in Table I. Clearly, designers, in recognizing the basic problems, are devoting considerable engineering effort towards solving them at a not inconsiderable cost penalty.

Reactor Systems

Clearly, a variety of solutions is available to the designer, any of which may be tailored to his particular needs, the ultimate choice being dictated by the particular design. However, a trend is discernible when comparing pressurized

6. The Reliability of Heat Removal Systems

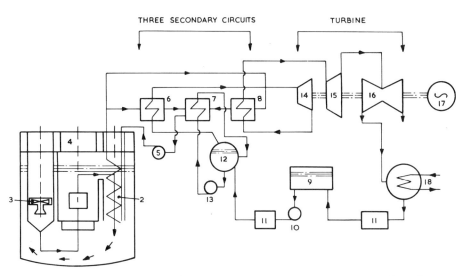

FIG.4 Plant flow diagram. 1, core and breeder; 2, six internal heat exchangers; 3, three primary pumps; 4, rotating shield; 5, secondary pump; 6, superheater; 7, evaporator; 8, reheater; 9, deaerator; 10, boiler feed pump; 11, feed heaters; 12, boiler drum; 13, circulating pump; 14, HP steam turbine; 15, IP steam turbine; 16, LP steam turbine; 17, generator; 18, condenser.

with unpressurized reactors because the depressurizing accident compounds the difficulties in arriving at a satisfactory solution. To illustrate this point, two reactor systems will be discussed.

The Prototype Fast Reactor

This is a sodium-cooled fast reactor of the pool type with a rated output of 250 MWe. The plant flow diagram is shown in Fig.4, and illustrates the reactor tank containing the primary circuits, the secondary sodium circuits, and the steam plant.

The primary vessel contains within it a second vessel called the reactor jacket, so that the primary sodium is separated into two pools—an inner pool and an outer pool. Sodium is drawn from the outer pool, pumped at about 100 psi into the plenum beneath the reactor core. It then passes through the core into the inner pool from which it falls by gravity through the intermediate heat exchangers back into the outer pool, thus completing the primary circuit.

There are six intermediate heat exchangers which are connected in pairs to three forced-circulation secondary sodium

TABLE I

Decay heat removal through steam circuits

Reactor	Heat sink redundancy	Power supply redundancy	Feed water redundancy
1	Main condenser; air cooling arrangements for SG	4 independent electrical sources	3 FW pumps
2	Main condenser; secondary sodium circuit air coolers (× 3)	—	—
3	Main condenser; blow off vents; loop specific water/ steam system (× 3)	All loops; specific heat removal systems have backup power supplies	3 main FW pumps; 3 emergency FW pumps
4	Main condenser; blow off vents	2 × turbo FW pumps; 1 × electric FW pump	3 FW pumps
5	Main condenser; blow off vents	FW pump diversity; main steam; electric, 10%; steam, 10%	3 FW pumps
6	Main condenser; blow off vents; 10% dump condenser	2 steam systems	
7	Main condenser; protected air-cooled condensers (× 3); blow off vents	2 FW pumps (electric motors); 1 FW pump (turbo driven); natural circulation capability	2 supplies with protected water tanks; 3 FW pumps
8	Main condenser; special condenser; blow off vents	All power from turbo generators; and diesels for emergency power supply	5 FW pumps and 2 emergency FW pumps
9	Main condenser; blow off vents; special condensers	Power from main turbo generators; emergency power from diesel	3 FW pumps to each steam generator; 3 emergency FW pumps with own power supply line

6. The Reliability of Heat Removal Systems

Various Approaches to Fast Reactor Decay Heat Removal Systems

	Decay heat removal from primary system sodium			
Location	Redundancy	Natural circulation	Main loop pony motor requirements	Remarks
Water cooling external to main reactor guard vessel, operating continuously	2 systems	Yes	Yes, plus flywheels	—
As above	—	Yes	Yes	—
Dip heat exchangers in reactor vessel	6 systems (3 systems have 100% capability)	No	—	—
Dip heat exchangers in primary IHX	3 systems	Yes, but not relied on	Yes	—
Dip heat exchangers in primary vessel (NaK)	3 systems each with 2 dip heat exchangers each 5 MW	Yes	Yes, part system continuing in operation	Failure criterion 650°C (bulk temperature)
Dip heat exchangers in primary vessel (NaK)	4 systems each with 2 dip heat exchangers each 13 MW	Yes; intended to cover pump rundown transients	Yes	As above
Main vessel sodium overflow system; part of system operates continuously	1 system, 2 air heat exchanger heat sinks	Yes, in each loop (possible short-term inadequacy)	Yes, to promote mixing for overflow heat removal	Failure criterion 1250°F (670°C) bulk temperature
None	—	—	—	—
None	—	—	—	—

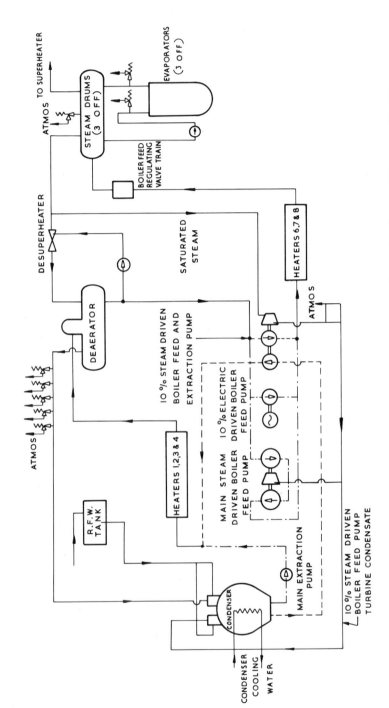

FIG. 5 *Decay heat rejection to condenser—normal electrical supplies available.*

6. The Reliability of Heat Removal Systems

circuits. A single secondary sodium circuit is illustrated, defining the flow path through the single-walled sodium heat exchangers. Heat is then passed to the steam generating units which raise steam for the single turbo alternator.

The main circuit parameters are:

Primary circuit
  Bulk sodium temperature   400°C
  Core outlet temperature   560°C
  Flow rate 23×10^6 lb/hr
Secondary circuit
  Sodium temperature at
    steam generator inlet   530°C
  Sodium temperature at
    steam generator outlet   370°C
  Flow rate   As for primary circuit

The normal heat rejection flow path may therefore be divided into three distinct parts: from the core to the intermediate heat exchangers, from the intermediate heat exchangers to the sodium water heat exchangers via the secondary sodium circuits, and then from the secondary sodium circuits to the steam system.

Under normal operating conditions, following a reactor trip, decay heat may also be rejected through this flow path using the main condenser as a dump condenser as shown in Fig.5. Boiler feed to the steam drums is maintained by either the 10% steam driven boiler feed pump and extraction pump or the main steam driven boiler feed pump and the main extraction pump. Because of the large heat rejection capability of this system, the average primary circuit operating temperature may be reduced to the reactor refueling temperature of 350°C in about half an hour and under normal circumstances will be used by the operator.

However, reliability analysis of the steam dump system showed that it would not meet the overall safety target for decay heat rejection by this method, as its probability of failure on demand was assessed to be in the 10^{-1}-10^{-2} bracket. It was judged to be uneconomic to improve the reliability to the standard required and an alternative system was installed comprising three thermal syphon loops with forced air cooling (see Fig.6). Each loop is completely independent and physically well segregated one from the other. The in-reactor heat exchange coils are contained within the intermediate heat exchangers where they receive a high degree of protection against any dynamic loadings from a reactor power excursion.

Hence, two separate and diverse decay heat rejection systems are provided, one of them the thermal syphon loops possessing considerable redundancy because one out of three is adequate in terms of capacity.

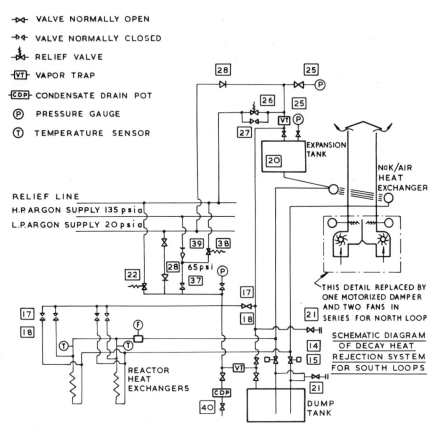

FIG.6 Flow diagram for decay heat rejection—circuit B.

Heat Removal in the Primary Circuits

This heat rejection path is common to both systems and so in itself must be shown to be adequate in relation to design duty and reliability. The total quantity of sodium in the primary system is about 1000 tons with a corresponding thermal capacity of 1150 MW-sec/°C which, together with the thermal capacity of the steel structure, amounts to some 1830 MW-sec/°C. Therefore, if after a trip all the decay heat were to be uniformly distributed in the primary sodium, it would only begin to boil in about 24 hr. This means that provided sufficient heat can be removed from the core and there is good mixing of the sodium in the outer tank, many hours are available for switching in external heat rejection loops.

Extracting heat from the core implies the need for either forced or natural circulation and if indeed the latter could be guaranteed at all times then that would be the solution.

6. The Reliability of Heat Removal Systems

Heat removal by natural circulation in the longer term is not in doubt. In the short term, however, immediately following a reactor trip the promotion of natural circulation in the transient phase and the consequences of flow stagnation, if indeed it occurred, are matters open to debate. This arises from the configuration of the primary circuit, where natural circulation requires only a head difference of a few inches between the inner and outer pools, whereas at power the head difference is measured in feet. Hence, during the primary pump rundown phase, until the levels adjust, natural circulation is inhibited. For these and other reasons forced circulation in the primary circuit is guaranteed by means of pony motor drives energized from the dc battery backed supplies and arranged to engage when pumps run down to 10% of nominal speed.

Reliability analysis of the pony motor drives has indicated an upper probability of a single pony motor failing to engage on demand of the order of 6×10^{-3} so that with three pony motors the probability of all drives failing is 2×10^{-7} per demand—a satisfactory result.

Decay Heat Rejection Capacity

The capacity of the steam dump system is considerably in excess of requirements and will bring primary circuit temperatures down very quickly. The limiting case is represented by the thermal syphon loops whose total capacity is 15 MW (5 MW per loop), and Fig.7 shows the core outlet temperature transients for one, two, or three loops invoked immediately following a reactor trip. The transients show that for two or three loops a core outlet temperature of 550°C is not exceeded but that for one loop the temperature peaks at 650°C. This is a plant design limit and so the criterion for system failure is the failure to start up one loop immediately on demand.

Reliability Analysis of Thermal Syphon Loops

The system considered comprises three independent heat rejection circuits, two of which are similar and illustrated in Fig.6. The third circuit is essentially the same except that the two motors and associated dampers operating in parallel are replaced by one motorized damper and two motors in series. Each circuit consists of two sodium/sodium-potassium alloy (Na/NaK) heat exchangers in the reactor tank connected in parallel to a single NaK/air heat exchanger by interconnecting pipework. In operation cooling is achieved by means

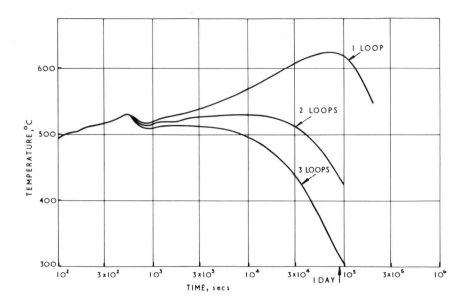

FIG.7 Core outlet temperature transients with 1, 2, or 3 DHR loops in operation immediately following reactor shutdown.

of natural circulation in the NaK system and forced draught cooling of the NaK/air heat exchangers by electrically operated fans. An expansion tank pressurized with an argon blanket gas to 45 lb/in.^{2} is illustrated together with a dump tank for emptying the circuit in an emergency. Pressure, temperature, and flow measurement stations are also shown.

Logic Diagram

The first stage in the analysis is to derive a logic diagram summarizing the requirements for successful operation and from which the failure probability may be calculated. The main logic diagram is shown in Fig.8.

Referring to Fig.8, which represents the failure modes for a single heat rejection circuit, entry is made at the start box on the top left-hand side terminating at the bottom right-hand side after following the direction of the arrows. Each box represents a failure mode for a particular component in the circuit which can be represented by a mathematical failure probability model; for example, box 20 in the bottom section represents the probability of expansion tank failure. By a summation process the individual probabilities are combined to estimate the failure probability of the whole circuit.

6. The Reliability of Heat Removal Systems

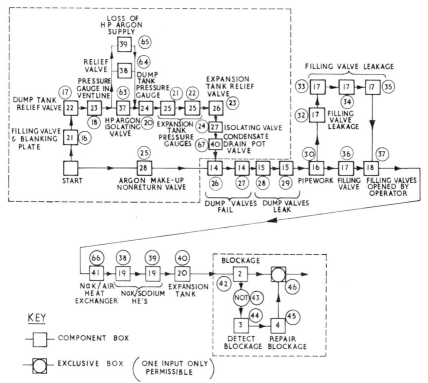

FIG.8 *Logic diagram for circuit subsystem. Numbers outside boxes refer to switch numbers in program; numbers in boxes refer to components listed in Table I.*

Three types of circuit failure are considered:

(a) loss of argon: boxes Start to 40, shown by dashed line enclosed section at top of Fig.8;
(b) loss of NaK: boxes 14 to 20; and
(c) circuit blockage: boxes 2 to 46, shown by dashed line enclosed section at bottom of Fig.8.

Taking one of these failure modes by way of example, loss of NaK can occur due to dump valves leaking or being inadvertently opened (boxes 14, 15), pipework fracture (box 16), filling valve failure (boxes 17, 18), rupture of the heat exchangers (boxes 41, 19), or expansion tank failure (box 20). The remaining failure modes are clear by inspection.

A single circuit is brought into operation by pressing the start button initiating the motor start sequence and damper operation. Failure of the contactor equipment or electric motor may be detected by a centrifugal switch on the motor,

TABLE II

Component	Type of failure	Consequence of failure
Start button	Failure to make contact	Natural circulation cooling only
Contactor	Fail to energize motor	Natural circulation cooling only available
Dump valves	Inadvertently opened	Rapid loss of sodium
	Leakage passed glands	Slow loss of sodium
Pipework (305 ft)	Split	Leakage of sodium into nitrogen-filled detection cavity
Filling vent valves (4 in parallel; 1 in series)	Leakage	Partial loss of sodium into dump tank
	Operator error	Partial loss of sodium into dump tank
HE coils in reactor (165 ft of 4-in. pipe)	Split	NaK leakage into reactor
Expansion tank	Failure	NaK fire
NaK/air heat exchanger	Leakage	NaK fire
Filling valve and blanking plate	Leakage passed gland and blanking plate	Leakage of argon cover gas
Dump tank relief valve set at 70 psia	Leaking valve	Leakage of argon cover gas
Pressure gauge in vent line	Leaking pressure gauge	Leakage of argon cover gas
Dump tank pressure gauge	Leaking pressure gauge	Leakage of argon cover gas
Expansion tank pressure gauges	Leaking pressure gauge	Leakage of argon cover gas
Expansion tank relief valve set at 25 psia	Leaking valve	Leakage of argon cover gas
Isolating valve in parallel with above	Seat leakage	Leakage of argon cover gas
Condensate drain pot valve	Seat leakage	Leakage of argon cover gas
Isolating valve joining gas line to 65 psi supply	Seat leakage	Pressurized argon gas blanket
Relief valve	Leaking seat	HP argon supply leak
Argon make-up non-return valve	Fails to open on loss of DHR argon pressure	Argon lost from DHR system if leak occurs
Motorized damper	Fail to open	Draught not available
Electric motor	Fail to start or run	Forced draught not available

6. The Reliability of Heat Removal Systems

Component Failure Rate and Repair Data—Dormant System

Fault rate (θr/yr)	Source of data	Failure probability to detect fault	Repair time while circuit is dormant
0.005	AHSB(S)R117	10^{-2}	2 hr
0.016		10^{-2}	4 hr
0.1	Assumed	10^{-1}	4 hr
0.1			3 days
3×10^{-6}	GEAP 4575 10^{-8} f/yr-ft	10^{-1}	10 days
10^{-2}	MI 60-54 Rev 1	10^{-1}	3 days
4×10^{-6}		10^{-1}	4 hr
1.6×10^{-5}	GEAP 4575 10^{-7} f/yr-ft	10^{-1}	1 month
10^{-4}	Assumed	10^{-1}	1 week
10^{-1}	Assumed	10^{-1}	1 month
5×10^{-6}	MI 60-54; 0.05 f/yr for valve leakage; 10^{-4} f/yr for blank plate failure	10^{-1}	4 hr
0.02	AHSB(S)R117	10^{-1}	4 hr
0.05	Valve leakage failure rate assumed	10^{-1}	4 hr
0.05	Valve leakage failure rate assumed	10^{-1}	4 hr
0.05	Valve leakage failure rate assumed	10^{-1}	4 hr
0.02	AHSB(S)R117	10^{-1}	4 hr
0.05	Valve leakage failure rate assumed	10^{-1}	4 hr
0.05	Valve leakage failure rate assumed	10^{-1}	4 hr
0.05	MI 60-54 Rev 1	10^{-1}	4 hr
0.02	AHSB(S)R117	10^{-1}	4 hr
0.02	Same figure assumed as for leaking valve	10^{-1}	4 hr
0.03	—	10^{-3}	$1 \pm 1/2$ hr
0.03	—	10^{-4}	1 month

enabling subsequent repair operations to rectify the defect. This repair facility will of course improve plant availability. Subsequently the success of the system will depend on the circuit being filled with NaK to the appropriate level, sealed off by argon, and adequate natural convection flow established with no blockage present. Gross circuit blockage can only be detected after a start signal which will promote natural circulation. Such a condition would be detected by the flow alarm (box 3, Fig.8).

The input data for circuit component failures are listed in Tables II and IV. Three types of fault have been considered:

(a) revealed faults which occur before a demand is placed on the DHR system;

(b) unrevealed faults occurring prior to a system demand; and

(c) faults occurring when the DHR system is running and which fall into the revealed category.

Hence, the data in the tables refer to revealed fault rates, the failure probability of fault detection equipment from which the unrevealed fault rate can be evaluated, and estimated repair times. Additionally, for those plant items which are not "on stream" regular proof testing is normally carried out so that the proof test period must be defined. In this particular example these items are the damper and motor start circuits. The assumptions made in the calculations are:

(a) revealed faults occurring before a system demand are continuously repaired. At the same time existing unrevealed faults are repaired.

(b) revealed faults occurring after a system demand are repaired; and

(c) air dampers remain in the open position after the system is operational.

(d) motor failures are repaired at all times, but it is noted that the mean repair time is as long as the time for which the system is required to run.

(e) proof tests are carried out on plant items at 50-day intervals.

Criterion for System Failure

Following a reactor trip, decay heat will decrease and there is a requirement to provide a minimum heat removal capacity. Clearly this value will vary with the time delay between reactor trip and making the DHR system operational. Increasing capacities will be required for increasing delay times, the

6. The Reliability of Heat Removal Systems

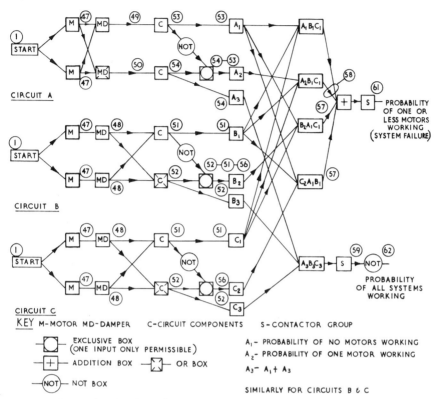

FIG.9 Main logic diagram—failure modes of decay heat rejection system. Numbers outside boxes refer to switch numbers in program.

limit being set by the permissible bulk sodium temperature rise in the tank. For the system under review the minimum heat removal capacity is achieved with any two motors operational within a 4-hr time scale. Subsequently, this minimum heat removal capacity must be maintained for 30 days. Clearly one is interested in the probability of system failure on demand and subsequently within a 30-day period.

Analysis Using NOTED Program

The analysis has been carried out using the NOTED program [19] which computes the time dependent failure probability for each block in the logic diagram and carries out the summation process. As any two motors operating in any circuit satisfies the basic criterion for success, it is necessary to compute the probability of one or less motors working. A further logic

TABLE III

System Failure Probability to Start for Various Fault Rates and Repair Times

Repair times	\multicolumn{6}{c}{Faults per year}					
	0.5	0.1	0.15	0.2	0.7	1.0
NaK/air heat exchanger						
1 day	1.08×10^{-5}	1.1×10^{-5}	—	1.17×10^{-5}	1.85×10^{-5}	2.56×10^{-5}
1 week	1.09×10^{-5}	1.16×10^{-5}	—	1.36×10^{-5}	3.93×10^{-5}	7.16×10^{-5}
1 month	1.19×10^{-5}	1.44×10^{-5} [a]	—	2.43×10^{-5}	2.03×10^{-4}	4.69×10^{-4}
3 months	1.3×10^{-5}	1.84×10^{-5}	—	4.06×10^{-5}	5.69×10^{-4}	1.39×10^{-5}
Dump valve						
4 hours	1.33×10^{-5}	1.44×10^{-5} [a]	1.58×10^{-5}	1.7×10^{-5}	—	7.5×10^{-5}
1 week	1.4×10^{-5}	1.63×10^{-5}	2.31×10^{-5}	2.8×10^{-5}	—	3.8×10^{-4}

[a]The figures shown represent the standard case defined by the data in Table I.

6. The Reliability of Heat Removal Systems

diagram is, therefore, required combining the various failure modes for all three circuits.

This is shown in Fig. 9, where M, D, and C represent the failure probabilities of the motor, damper, and circuit subsystems, respectively. Taking circuit A, the box A_1 represents the failure probability of the circuit or both motors or damper. Box A_2 is the probability of one motor working and circuit A healthy. Similarly for circuits B and C.

Hence the probability of no motors working is A, B, C and the probability of one motor working is $A_2B_1C_1 + B_2A_1C_1 + C_2A_1B_1$. The sum total $A_1B_1C_1 + A_2B_1C_1 + B_2A_1C_1 + C_2A_1B_1$ then represents the probability of one or less motors working, i.e., system failure.

Results

The analysis has been considered in two separate phases:

(a) the probability of system failure on demand at any time having regard to the continuous failure and repair process for revealed faults, and

(b) the probability of failure in operations subsequent to a demand, when all faults will be revealed.

In the latter event there are a large number of starting conditions as the system may be in a large number of partially failed states on demand.

The standard case may be defined as that representing the failure rates and repair times listed in Table II. For this case the system failure probability on demand immediately prior to the 50-day proof test is 1.44×10^{-5}. How this failure probability varies with different assumptions regarding two sensitive items, namely, the NaK/air heat exchangers and dump valves, is illustrated in Table III and plotted in Figs.10 and 11.

For small repair times Fig.10 demonstrates that the system failure probability is a logarithmic function of the fault rates for the NaK/air heat exchangers. Clearly to achieve a reliability in the 10^{-4}-10^{-5} band, the fault rate must be limited to 0.3 f/yr if the repair time is as long as 3 months. On the other hand, if repair can be effected within one week, a fault rate up to 1.0 f/yr is acceptable.

It is interesting to note from Fig.11 that much more stringent requirements are needed to meet the 10^{-4}-10^{-5} target for failures associated with the dump valves. For fault rates up to 1 f/yr the repair time has to be limited to 4 hr, or if 1 week is assumed for repair the maximum acceptable fault rate is 0.6 f/yr. The reason for this is quite simple; it is because there are two dump valves per circuit, either of which

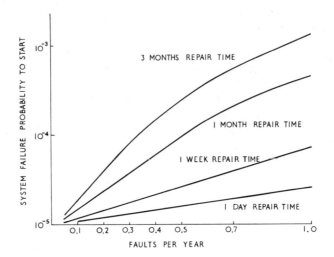

FIG.10 DHR system failure probability for varying fault rates and repair times for the NaK/air heat exchangers.

can render the circuit inoperative if the valve leaks or is inadvertently opened by an operator.

While the system is running, all faults will be revealed as a loss of heat extraction rate. The system failure probability in service has then been computed in accordance with the assumptions listed in Table IV.

The two sets of results relating to failure probability on demand and in service may now be combined to assess the combined failure probability. Let P be the probability of system failure to start and meet the design criterion, Q the probability that the system started perfectly, p the probability of failure in service to meet the design criterion assuming the system started perfectly, and, q the probability that the system runs perfectly after starting perfectly. Then

combined failure probability = probability of system being in
the failed state at start-up
+ probability of failing while
in service
= P + Qp

Note that in this analysis all the intermediate states between system failure at start-up and a perfect system at start-up have been ignored. The value of P, as seen from Figs.10 and 11, is in the 10^{-3}-10^{-5} band depending on the assumptions made. The value for Qp is sensibly constant over the 30-day period and is also much less than the failure probability on demand. Hence, it is evident that the results shown in Figs.10 and 11

6. The Reliability of Heat Removal Systems

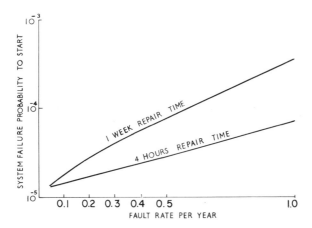

FIG.11 DHR system failure probability for varying fault rates and repair times for the dump valves.

may be regarded as the system failure probability in service because the failure probability while running is an order of magnitude lower than the failure probability to start. This is because the system is only required to run for about 30 days, whereas the failure probability to start is averaged over the 30-yr life involving a continuous failure and repair process which for the most sensitive component may vary between 1 and 90 days.

One may, therefore, conclude that a failure probability of 10^{-5} per demand is achievable taking realistic values for fault rates and repair times. Combining the steam dump system with the thermal syphon loops suggests a failure probability per demand of 10^{-7}. Based then on a demand frequency of 10/yr, the frequency of a whole core meltdown arising from failure to reject decay heat is estimated to be 10^{-6}/yr.

Common Fault Modes

Common fault modes arising either from external or reactor accidents negate all redundancy arguments based on statistical independence of randomly occurring faults, and so it is necessary to consider them. As the reactor is operating at approximately atmospheric pressure, depressurizing accidents cannot occur. However, the loss-of-coolant accident represents a hypothetical situation which must be considered.

The primary sodium is contained in a double container, namely, an inner primary tank and an outer leak jacket. A leakage in the primary tank would result in a lowering of the

TABLE IV

Component	Type of failure	Consequence of failure
Start button	Failure to make contact	Natural circulation cooling only
Contactor	Fail to energize motor	Natural circulation cooling only available
Dump valves	Inadvertently opened	Rapid loss of sodium
	Leakage passed glands	Slow loss of sodium
Pipework (305 ft)	Split	Leakage of sodium into nitrogen-filled detection cavity
Filling vent valves (4 in parallel; 1 in series)	Leakage	Partial loss of sodium into dump tank
	Operator error	Partial loss of sodium into dump tank
HE coils in reactor (165 ft of 4-in. pipe)	Split	NaK leakage into reactor
Expansion tank	Failure	NaK fire
NaK/air heat exchanger	Leakage	NaK fire
Filling valve and blanking plate	Leakage passed gland and blanking plate	Leakage of argon cover gas
Dump tank relief valve set at 70 psia	Leaking valve	Leakage of argon cover gas
Pressure gauge in vent line	Leaking pressure gauge	Leakage of argon cover gas
Dump tank pressure gauge	Leaking pressure gauge	Leakage of argon cover gas
Expansion tank pressure gauges	Leaking pressure gauge	Leakage of argon cover gas
Expansion tank relief valve set at 25 psia	Leaking valve	Leakage of argon cover gas
Isolating valve in parallel with above	Seat leakage	Leakage of argon cover gas
Condensate drain pot valve	Seat leakage	Leakage of argon cover gas
Isolating valve joining gas line to 65 psi supply	Seat leakage	Pressurized argon gas blanket
Relief valve	Leaking seat	HP argon supply leak
Argon make-up non-return valve	Fails to open on loss of DHR argon pressure	Argon lost from DHR system if leak occurs
Motorized damper	Fail to open	Draught not available
Electric motor	Fail to start or run	Forced draught not available
Pipework	Blockage	Limited cooling

6. The Reliability of Heat Removal Systems

Component Failure Rate and Repair Data—Running System

Fault rate ($\theta r/yr$)	Source of data	Failure probability to detect fault	Repair time while circuit is operational
Zero	AHSB(S)R117	Zero	Not applicable
0.016		Zero	2 hr
0.1	Assumed	Zero	4 hr
0.1			3 days
3×10^{-6}	GEAP 4575 10^{-8} f/yr-ft	Zero	10 days
10^{-2}	MI 60-54 Rev 1	Zero	3 days
4×10^{-6}		Zero	4 hr
1.6×10^{-5}	GEAP 4575 10^{-7} f/yr-ft	Zero	1 month
10^{-4}	Assumed	Zero	1 week
10^{-1}	Assumed	Zero	1 month
5×10^{-6}	MI 60-54; 0.05 f/yr for valve leakage; 10^{-4} f/yr for blank plate failure	Zero	4 hr
0.02	AHSB(S)R117	Zero	4 hr
0.05	Valve leakage failure rate assumed	Zero	4 hr
0.05	Valve leakage failure rate assumed	Zero	4 hr
0.05	Valve leakage failure rate assumed	Zero	4 hr
0.02	AHSB(S)R117	Zero	4 hr
0.05	Valve leakage failure rate assumed	Zero	4 hr
0.05	Valve leakage failure rate assumed	Zero	4 hr
0.05	MI 60-54 Rev 1	Zero	4 hr
0.02	AHSB(S)R117	Zero	4 hr
0.02	Same figure assumed as for leaking valve	Zero	4 hr
Zero	—	Zero	Not applicable
0.03	—	Zero	1 month
0.01	Assumed	10^{-2}	180 hr

sodium levels in the primary tank, the extent of the leakage being determined by the interspace volume. In the short term, the decay heat rejection coils would remain immersed and would function normally. However, in the longer term partial uncovering of the coils might occur due to sodium absorption by insulation and this would lead to a reduction in decay heat rejection efficiency. This can be readily accommodated by topping up the system, there being adequate time available to do this. In any case, because the primary tank is operating at 400°C, the critical crack length is very large and catastrophic failure unlikely. Judgment, therefore, rests on a leak before failure mode, in which case leakage detection monitors would alert the operator to the situation so that appropriate emergency procedures could be readily instituted.

Common mode faults arising either from mechanical damage or failure of power sources have also been considered. These include collapse of secondary containment building, damage due to pump flywheel fracture, damage due to dropped crane loads, propagation of fire initiated in secondary sodium circuits, and failure of dc/ac electrical supplies.

Pressurized Reactors

With pressurized reactors a new dimension is added to the emergency core cooling problem because of the reduced capacity of the coolant following depressurizing accidents. The difficulties are further compounded when comparing pressurized water reactors with gas-cooled reactors because of the phase change as the water flashes off to steam during depressurization, thus decreasing the heat removal efficiency.

The situation with respect to gas-cooled reactors is discussed in Chapter 8, where the solution is seen in terms of limiting the frequency and rate of depressurization by detailed design attention to the closures of the prestressed concrete pressure vessel. Hence, the special design characteristics of the prestressed concrete pressure vessel are used to advantage in allowing a grace period following which forced cooling can be instituted using the diesel-backed electrical supplies as a power source. For nondepressurizing accidents auxiliary cooling has to be provided to high standards of reliability.

The solutions for PWRs as reflected in US experience are well exemplified in the approach illustrated in Ref.[35], to which the reader is referred. The plant considered is the PWR Surry Power Station, Unit 1, 788 MWe capacity.

With the reactor pressurized, a number of accident states have been evaluated, the limiting case being the loss of offsite power. One of the consequences of this initiating event

6. The Reliability of Heat Removal Systems

is the fact that the steam main feed train becomes inoperable, in which case reliance has to be placed in the auxiliary feed water system. Two reserve feed water tanks are provided together with a connection to the fire main providing copious water supplies. Three pumps, two electrical and one steam-driven, are provided, any of which is of adequate capacity to deliver water to the steam generator which can discharge steam to the atmosphere through motorized valves. It will, therefore, be appreciated that there is considerable system redundancy. Following the loss of off-site power (it is presumed that natural circulation in the primary circuit is adequate following primary pump rundown), it is only a question of removing decay heat from the steam generators. The auxiliary feed water system performs this function and is started automatically following loss of external electrical supplies.

The loss-of-coolant accident frequency is then defined as the product term TMLB'ε (using the notation in the report), where

$T = 2 \times 10^{-1}$/yr = initiating event frequency (loss of external electrical supplies)

$M = 2 \times 10^{-1}$ = probability of failing to restore external supplies within 1/2 hr and represents probability of loss of main feed

$L = 1.5 \times 10^{-4}$ = probability of failure of auxiliary feed water system

$B' = 5 \times 10^{-1}$ = probability of failure to recover either on or off-site power within 1 1/2 hr

$\varepsilon = 8 \times 10^{-1}$ = probability of melt through the base of the vessel

The loss-of-coolant accident frequency is then computed as 2×10^{-6}/yr. It is instructive to note that a highly reliable cooling system is provided to cover the contingency of an accident with the reactor pressurized, representing a convergence of US/UK judgment. Furthermore, because the core meltdown proceeds through the base of the tank, only a small fraction of core activity is released so that the reliability target for the cooling system is justified.

The loss-of-coolant accident in pressurized water reactors presents formidable problems [30] in gaining confidence that the engineered safeguards are adequate. However, the lack of confidence stems not from difficulties in providing hardware having the requisite reliability, but rather from a lack of complete understanding of the basic heat removal processes in the complex hydrodynamics following a depressurizing accident. Four types of circuit fracture are considered with the following accident frequencies:

Pressure vessel failure	10^{-7}/yr
Pipe rupture equivalent to 6-in. diameter hole	10^{-4}/yr
Pipe rupture between 2 and 6-in. equivalent diameter	3×10^{-4}/yr
Pipe rupture between 1/2 and 2-in. equivalent diameter	10^{-3}/yr

Because of the very low frequency of pressure vessel failure, the question of availability of emergency core cooling does not arise. For the pipe rupture accidents, ECCS failure probabilities between 10^{-2} and 10^{-3} per demand have been evaluated as being satisfactory. In this context the ECC system comprises four subsystems, namely, the accumulator, low-pressure injection, high-pressure injection, and safety injection control systems. Evidently the provision of ECC hardware to the required reliability standards does not prove to be an intractable problem.

Discussion

The application of probabilistic theory to the quantitative evaluation of reactor emergency cooling systems is rewarding as a means for rationalizing the required strategy. The procedure requires the setting of risk targets which invariably embody the concept that low-hazard events are acceptable on a comparatively high-frequency basis when compared with large hazards.

As applied to emergency core cooling systems, a prime requirement is the provision of high-reliability systems for normal operating conditions. Not only must the systems be available on demand with a low-failure probability, but there must also be considerable confidence that they will work. In respect to performance there are two distinct operational phases which merit attention. First of all, there is the transient phase immediately following reactor trip, during which core flow stagnation must not occur. Then there is the question of long-term heat removal, where reliance may be placed on redundant systems and where common mode faults due to failure of electrical or water supplies assume some importance. This author believes that simple systems employing the minimum number of components where reliance can be placed on natural circulation is the most effective and desirable solution. The examples quoted in the text show that the risk targets can be met provided proper consideration is given to achieving them at the design stage.

6. The Reliability of Heat Removal Systems

Appendix. Mathematical Models

Statistical Failure Distributions

The failure rate for any device will, in general, be a function of its age T. Thus the probability that the device will fail during the interval when its age increases from T to T + dt is given by the expression θ(T) dT. If the device is not subject to repair or maintenance and is such that once it has failed it remains failed forevermore, the probability P(T) that it would be found failed when it is of age T is given by the differential equation

$$dP(T)/dT = [1-P(T)]\theta(T) \tag{1}$$

This can be integrated to give

$$P(T) = 1-\exp[-\int_0^T \theta(t)\, dt] \tag{2}$$

which is the mathematical expression defining the reliability characteristics for the device in this particular case.

In general, we can postulate a function P expressing the probability that the device will be found failed at the time it may be called upon to operate, but this function will normally depend on the past history of the device as well as its age. It is possible, however, in most practical cases to divide the lifetime of a device into intervals such that either P maintains a constant value through the interval, or P is given by an expression of the form of Eq.(2) with T taking the value zero at the beginning of the interval. The first case would be appropriate for, say, starting a diesel generator when the probability that it will start when called upon is assumed not to depend on whether or not it started when tried on a previous occasion. The second case would be appropriate, for example, to an electric light bulb which is periodically replaced, T being the number of hours it had been burning since last replaced.

Expression (2) for any particular device can only be obtained by experiment, or in an equivalent way, which will either give estimates of P(T) at a set of discrete values of T or will give values of "age to failure" for a number of devices which have failed. As a result, there is rarely enough data to derive directly the form of expression (2). It is better to choose a small number of forms for Eq.(2) depending on one or two parameters and find the one that most closely fits the observed data.

One- and Two-Parameter Distributions

Of one-parameter forms it is usual only to consider the exponential distribution obtained by making the failure rate $\theta(T)$ a constant independent of T. This gives

$$P(T) = 1-\exp(-ST) \tag{3}$$

which is an expression that fits tolerably well to many devices through most of their lifetime.

Four types of two-parameter forms are suitable. These are:

Log normal:

$$P(T) = (S\sqrt{2\pi})^{-1} \int_{-\infty}^{\log T} \exp[-(x-R)^2/2S^2]\, dx \tag{4}$$

Normal:

$$P(T) = (S\sqrt{2\pi})^{-1} \int_{-\infty}^{T} \exp[-(x-R)^2/2S^2]\, dx \tag{5}$$

Special Erlangian:

$$P(T) = 1 - \left[1+RT+\frac{(RT)^2}{2!}+\cdots+\frac{(RT)^{S-1}}{(S-1)!}\right]\exp(-RT) \tag{6}$$

Weibull:

$$P(T) = 1-\exp(-T^S/R) \tag{7}$$

Of these the Weibull is particularly attractive as it reduces to the exponential if $S = 1$ and can approximate fairly closely to the other distributions by suitable choice of the two parameters R and S.

When the device is subject to repair or maintenance, these simple expressions will be modified. To avoid confusion in the following development, we will use $Q(T)$ to represent one of the probability expressions given in Eqs.(3)-(7).

Regular Maintenance

For a device which is taken out of service at regular intervals I for maintenance, being returned after a further interval J, fully serviced, define $t = T \bmod(I+J)$. Then

$$\begin{aligned} P(T) &= Q(t) \quad \text{if } t < I \\ &= 1 \quad \text{if } t \geq I \end{aligned} \tag{8}$$

since the device cannot operate if called upon when it is out of service for maintenance.

Regular Inspection

A variation of maintenance schedule is to assume inspection at regular intervals, the device being replaced if faulty. For simplicity, assume that the inspection and replacement takes place instantaneously, i.e., the J used in the previous paragraph is zero. The values of P(T) are now given by the series of equations

$$P_i(T) = P_{i-1}(T) - P_{i-1}(iI)[1 - P_0(T-iI)] \quad i = 1, n$$

$$P_0(t) = Q(t) \quad \text{for all } t \tag{9}$$

$$P(T) = P_n(T)$$

where n is the largest integer such that $nI < T$. The notation used is such that $P_i(t)$ represents the probability that the device would be found failed at time t having had i inspections.

If Q(t) is of the exponential form given by Eq.(3), it can be readily verified that

$$P(T) = P_n(T) = 1 - \exp[-S(T-nI)] = Q(T-nI)$$

showing that P(T) depends only on the time that has elapsed since the last inspection. This is precisely the result that would arise from Eq.(8) in similar circumstances (i.e., J = 0). This simplification is peculiar to the exponential law and can be shown to occur only in this case.

Immediate Repairs

Any device is subject to two types of fault. Revealed faults become immediately apparent and can be repaired without delay; unrevealed faults do not show themselves until the device is called upon to operate. It is reasonable to assume that unrevealed faults are corrected during a maintenance period or when a revealed fault is repaired.

On these assumptions suppose that revealed faults occur according to the law $Q_r(T)$ and unrevealed faults according to the law $Q_u(T)$. On occurrence of a revealed fault, suppose further that repair starts immediately and R(t) is the probability that repair is complete within a time t, R(t) being of one of the types given in Eqs.(3)-(7).

The probability that a revealed fault occurs between time t and time t+dt after the last repair is

$$q_r(t) = dQ_r(t)/dt$$

and the probability that a repair is completed between time t and time t+dt after the revealed fault occurs is

$$r(t) = dR(t)/dt$$

The function H(t), defined by

$$H(t) = \int_0^t q_r(y) r(t-y) \, dy \tag{10}$$

is the probability that a repair is completed at a time between t and t+dt after the previous repair was completed or after the device was new. So

$$F(T) = H(T) + \int_0^T F(u) H(T-u) \, du \tag{11}$$

is the probability that a repair is completed on the unit when it is of age between T and T+dT.

The probability P(T) that the unit is found failed at age T is then made up of the probability that a fault, either revealed or unrevealed, has occurred at some time before T less the probability that a repair had been completed at some time before T and no failure had occurred since. So the expression for P(T) is

$$P(T) = Q_r(T) + Q_u(T) - Q_r(T) Q_u(T)$$

$$- \int_0^T F(y) [1 - Q_r(T-y)] [1 - Q_u(T-y)] \, dy \tag{12}$$

The three, Eqs.(10), (11), and (12), together can be solved for P(T). In general, they cannot be solved analytically and their numerical evaluation involves what is effectively a triple integration. In the special case for which the functions $Q_r(t)$, $Q_u(t)$, and R(t) are all of the exponential type of Eq.(3), they can be solved analytically by means of Laplace transforms, giving

$$P(T) = A - [1 - Q_r(T)] \{B[1 - Q_u(T)] + C[1 - R(t)]\} \tag{13}$$

where

$$A = \frac{\lambda}{\lambda + \tau} + \frac{\mu \tau}{(\mu + \lambda)(\lambda + \tau)}, \quad B = 1 - \frac{\lambda \tau}{(\tau - \mu)(\mu + \lambda)}, \quad C = \frac{\lambda \tau}{(\tau - \mu)(\lambda + \tau)}$$

λ, μ, and τ being the constants occurring in the formulas for $Q_r(t)$, $Q_u(t)$, R(t), i.e.,

$$Q_r(t) = 1 - \exp(-\lambda t) \quad Q_u(t) = 1 - \exp(-\mu t) \quad R(t) = 1 - \exp(-\tau t)$$

It will be seen from Eq.(13) that as T becomes large, P(T) approaches the constant value A, i.e.,

$$P(\infty) = \frac{\lambda}{\lambda + \tau} + \frac{\mu \tau}{(\mu + \lambda)(\lambda + \tau)} \tag{14}$$

This can be shown to be a general result even if the initial distributions are not of the exponential type if $1/\lambda$ is interpreted as the mean time to failure by a revealed fault, $1/\mu$

6. The Reliability of Heat Removal Systems

is interpreted as the mean time to failure by an unrevealed fault, and $1/\tau$ is interpreted as the mean time for repair. The device thus settles into an equilibrium state with a constant failure probability given by Eq.(14).

References

1. Farmer, F. R. Siting Criteria—A New Approach, IAEA Symposium on the Containment and Siting of Nuclear Power Reactors, Vienna, April, 1967, Paper No. SM-89/34.
2. Bazovsky, I. "Reliability Theory and Practice." Prentice-Hall, Englewood Cliffs, New Jersey, 1961.
3. Ireson, W. G. "Reliability Handbook." McGraw-Hill, New York, 1966.
4. Pieruschka, E. "Principles of Reliability." Prentice-Hall, Englewood Cliffs, New Jersey, 1963.
5. Shooman, M. L. "Probabilistic Reliability; An Engineering Approach." McGraw-Hill, New York, 1968.
6. Barlow, R. E., and Proschan, F. "Mathematical Theory of Reliability." Wiley, New York, 1965.
7. Green, A. E., and Bourne, J. "Reliability Technology." Wiley, New York, 1972.
8. Amstadter, B. L. "Reliability Mathematics; Fundamentals, Practices, Procedures." McGraw-Hill, New York, 1971.
9. Myers, R. H., Wong, K. L., and Gordy, H. M. "Reliability Engineering for Electronic Systems." Wiley, New York, 1964.
10. Dummer, G. W. A., and Griffin, N. "Electronics Reliability Calculation and Design." Pergammon Press, Oxford, 1966.
11. Sandler, G. H. "System Reliability Engineering." Prentice-Hall, Englewood Cliffs, New Jersey, 1963.
12. Aeronautical Radio, Inc. "Reliability Engineering." Prentice-Hall, Englewood Cliffs, New Jersey, 1964.
13. Calabro, S. R. "Reliability Principles and Practices." McGraw-Hill, New York, 1962.
14. Guedenko, B. V., Belyayev, Yu. K., and Solovyev, A. D. "Mathematical Methods of Reliability Theory." Academic Press, New York, London, 1970.
15. Roberts, N. H. "Mathematical Methods in Reliability Engineering." McGraw-Hill, New York, 1964.
16. Chorafas, D. N. "Statistical Processes and Reliability Engineering." van Nostrand, Princeton, New Jersey, 1960.
17. Polovko, A. M. "Fundamentals of Reliability Theory." Academic Press, New York, London, 1968.
18. Zelen, M. "Statistical Theory of Reliability." Univ. of Wisconsin Press, Madison, Wisconsin, 1963.

19. Woodcock, E. R. The Calculation of Reliability of Systems. The Programme NOTED. UKAEA Rep. No. AHSB(S)R117 (1968).
20. Colombo, A. G. CADI: A Computer Code for System Availability and Reliability Evaluation, European Atomic Energy Community, EUR4940e (1973).
21. Veseley, W. E. A time-dependent methodology for fault-tree evaluation, *Nucl. Eng. Des.* 13, 337-360 (1970).
22. Colombo, A. G., Ricchena, R., and Volta, G. Survey and critical analysis of programmes for system reliability computation, Crest meeting MRR90, 1971.
23. Veseley, W. E., and Narum, R. E. Prep. and Kitt: Computer Codes for the Automatic Evaluation of a Fault-Tree, Idaho Nuclear Corporation, Rep. No. IN-1349 (August,1970).
24. Fussel, J. B. Synthetic Tree Model, A Formal Methodology for Fault Tree Construction, Aerojet Nuclear Company, ANCR-1098 (1911).
25. Morison, W. G., Pickering generating station, *Chart. Mech. Eng.* 62-66 (July 1975).
26. AEC-tr-7130. Phenix, Prototype Fast Neutron Nuclear Power Station, transl. of selected articles from *Bull. Inform. Sci. Tech.* (Paris) No. 138 (June 1969).
27. Steam generating and other heavy water reactors, BNES Conference Papers, May 14-16, 1968.
28. European Association for Gas Breeder Reactors, Brussels, GBR4 Safety Working Document (April 1975).
29. Rippon, S. The Rasmussen study on reactor safety, *Nucl. Eng.* (Dec. 1974).
30. Report to the APS by the study group on light water reactor safety, *Rev. Mod. Phys.* 47, Suppl. 1 (1975).
31. Alexander, T. The big blowup over nuclear blow drums, *Fortune 500* (May 1972).
32. Balfanz, H. P., Henser, F. W., and Ullrich, W. Principles of reliability analysis methods applied to emergency core systems, *Nucl. Eng. Des.* 29, No. 3, 384-394 (1974).
33. Boehm, B. Non-availability investigations of emergency core cooling systems (assessment of repair and inspection strategies), *Atomwirtsch. Atomtech.* 19, No. 7, 368-372 (July 1974).
34. USAEC. Acceptance Criteria for Emergency Core Cooling Systems for Light Water Cooled Reactors, Washington, D. C., Docket No. RM-50-1 (1973).
35. USAEC. An assessment of accident risks in U.S. commercial nuclear power plant (Rasmussen report), WASH-1400 (Aug. 1974).
36. Hoertner, H. Problems of reliability analysis in nuclear power plant technology, Second Information Meeting of the Laboratorium fuer Reaktorregelung und Anlagensicherung

6. The Reliability of Heat Removal Systems

Garching, Germany (Jan. 16, 1974).
37. Basil, W., and Hoertner, H. Reliability analysis of a PWR emergency core cooling system, Proceedings of the CREST Specialist Meeting on Applicability of Quantitative Reliability Analysis of Complex Systems and Nuclear Plants in Its Relation to Safety Held in Munich, Germany, May 26-28, 1971, Paper 3, Section III.
38. Basil, W., and Hoertner, H. Reliability analysis of a PWR emergency core cooling system. Continuation of investigation of a PWR emergency core cooling system, CREST Specialist Meeting held in Munich, Germany, May 26, 1971.
39. Bordelon, F. M., Massie, H. W., Jr., and Zordan, T. A. Westinghouse Emergency Core Cooling System Evaluation Model, Summary WCAP-8339 (June 1974).
40. Balfanz, H. P., Heuser, F. W., and Ullrich, W. Reliability approach for inspection strategy applied to an emergency core cooling system, Second International Conference on Structural Mechanics in Reactor Technology, Berlin, Sept. 10, 1973.
41. Karwat, H. Carrying Out of Realistic Calculations as a Basis for the Design of Emergency Core Cooling Systems, Technische Univ. Muenchen, Garching, Germany, Lab. fuer Reaktorregelung und Anlagensicherung. AED-CONF-73-500-2 (1973).
42. Heuser, F. W. Methodical Possibilities of Reliability Analysis for the Evaluation of Emergency Core Cooling Systems, Institut fuer Reaktorsicherheit der Technischen Ueberwachungs-Vereine e.V, Koeln, Germany, AED-CONF-73-500-3 (1973).
43. Balfanz, H. P. Statements of Applied Reliability Analysis, Institut fuer Reaktorsicherheit der Technischen Ueberwachungs-Vereine e.V, Koeln, Germany, AED-CONF-73-500-4 (1973).
44. Cottrell, W. B. Selected Bibliography on Emergency Core Cooling Systems for Light Water Cooled Power Reactors, ORNL-NSIC-113 (1974).
45. Dahll, G., and Hansen, O. Reliability Techniques: Data, Methods, Application, Institutt for Skipsmaskineri, Oslo, Norway, DRAM-9 (1973).
46. Wolf, L. Emergency cooling in nuclear power plants, *Energie 26,* No. 3, 119-121 (March 1974).
47. Farber, G. Problems in the evaluation of emergency core cooling, *Tech. Ueberwach. 15,* No. 3, 88-92 (March 1974).
48. Wilson, R. The AEC and the loss of coolant accident, *Nature 241,* No. 5388 (Feb. 1973).

7

The Integrity of Pressure Vessels
R. O'Neil

Introduction

The remarks in this chapter are confined to steel pressure vessels typical of those used in light water reactors. With the important exception of reinforced concrete pressure vessels, very little generality is lost by this restriction and the principles outlined are applicable to the majority of pressure vessels in the world's nuclear power plants.

In considering the safety of nuclear plants, the concept of a risk/consequence relationship has been introduced. It is reasoned that the permissible probability of encountering a given hazard can and should be related to the consequences of that hazard.

The isotope of iodine, ^{131}I, would usually provide an important external hazard to population immediately following pressure vessel failure, and may conveniently be used to express a quantitative measure of consequence. The amount of ^{131}I (in curies) released to the atmosphere may thus be related to the maximum permissible frequency of the release.

Figure 1 of Chapter 4 represents a proposed release/frequency limit for land-based power reactors [1], and will be used as a basis against which to evaluate the requirements for pressure vessel integrity in this chapter.

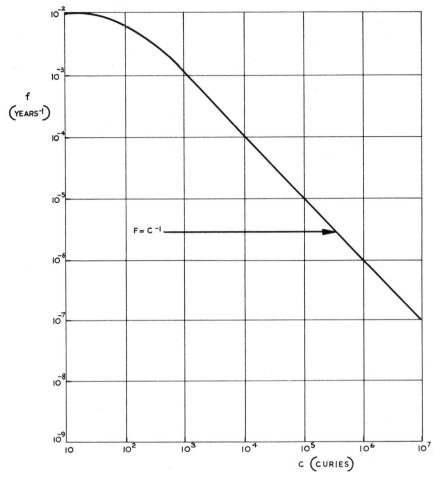

FIG.1 *Release frequency limit lines.*

Requirements

Figure 1 of Chapter 4 represents proposed limits for atmospheric release of ^{131}I from all accidents. However, failure of the pressure vessel is a dominating mode, and for major accidents the limits of Fig. 1 may be applied singly and directly for the purposes of this paper. Since the requirements relate to release to atmosphere, the role of all "postfailure" safeguards must be explicitly considered for any specific proposal. However, of these the integrity of the containment structure is probably the most important at the large accident end of the spectrum.

7. The Integrity of Pressure Vessels

The ways in which a pressure vessel might fail have been vigorously discussed for many years. There are different views about the possibility of catastrophic failure, fast fracture, brittle fracture, missile generation, etc. However, these detailed considerations can be encompassed within overall integrity arguments as developed in this section.

To proceed with a simplified argument, three categories of failure are considered.

Category A. Catastrophic failure of the main pressure vessel with fragmentation and/or pressure pulses, capable of severely damaging the containment.

Category B. Gross ductile rupture of the main pressure vessel with or without fragmentation, but limited either by the nature of the defect or surrounding structures to the point where the containment structure is not at risk, although the leakage rate may be increased.

Category C. Small-scale rupture or limited leakage within the design capacity of the containment.

It is important to realize that the safety and reliability assessment of a complete reactor system endeavors to embrace all possibilities of failure, not only of the main reactor pressure vessel but of a number of subsidiary vessels, pumps, valves, pipes, and other components. These are related to the reliability and capacity of emergency cooling devices, make-up, and flooding systems, containment sprays and filters, containment, melt-through protection, etc. While such an analysis is outside the scope of this chapter, the categories proposed and supported by simplified arguments are typical of those met in practice.

If we now consider failures of the types A, B, and C, it is clear by definition that the consequences of A are greater than B which, in turn, are greater than C. If we had sufficient experience to assess the likelihood of occurrence of pressure vessel failure, then clearly we would set a lower target for A than B, and lower for B than C.

Consider A—a catastrophic failure with damage to the containment. It is unlikely that all volatile fission products will escape, even from a damaged containment; the release to the atmosphere is reduced by partial retention in fuel, some absorption on surface, or washout in condensed steam or by water droplet. In some cases the consequences may be lessened by thermal uplift induced by self heating of the fission product cloud. The release will seldom exceed 10^7 Ci ^{131}I and may often be far less—on balance, our judgment is that the target of Fig.1 requires a category A failure not to exceed 10^{-6} per vessel per year. Category B failures causing only limited damage to the containment might be set at 10^{-5}

TABLE I

Summary of Pressure Vessel Reliability Requirements

Category	Vessel state	Reliability requirement (per vessel year)
A	Gross failure of the main pressure vessel coupled with significant fragmentation. May severely damage containment structure	10^{-6}
B	Gross failure of the main pressure vessel, without significant fragmentation. May significantly increase containment leakage rate	10^{-5}
C	Small-scale rupture within the design capacity of the containment	10^{-3}-10^{-4}

per vessel per year, whereas category C falls within the design capacity of the safeguards and might be expected to lead to an escape to the atmosphere, as with the design basis accident [2] of about 100 Ci ^{131}I.

Having in mind the consequent loss of confidence, cost, etc., the target might be set at 10^{-3}-10^{-4} per vessel per year. This is lower than might be determined by the direct consideration of possible fatalities, the risk of which should be minimal.

Statistical Evidence

Three approaches are used to consider the likelihood that a reactor pressure vessel designed, constructed, operated, and inspected to the latest standard can meet the requirements of Table I, viz.:

(i) Consideration of operating experience from all available commercial and military nuclear reactor pressure vessels.

(ii) Consideration of operating experience of all relevant nonnuclear pressure vessels with an appraisal of the differences between nuclear and nonnuclear vessels in respect of design, materials, manufacture, operation, and inspection.

(iii) Synthesis of failure probability using a model where the estimated residual critical defect is compared with the calculated critical defect having regard to the likely scatter of material properties. Such an analysis might be extended to take into account the probability of encountering specific transients and variability in crack growth data.

7. The Integrity of Pressure Vessels

TABLE II

Failures Found in Service in UK Pressure Vessel Survey[a]

	1962-1967		1967-1972	
Method of identification	Number of cases	%	Number of cases	%
Potentially dangerous failures				
Visual examination	75	56.8	34	24.5
Leakage	38	28.8	49	35.3
Nondestructive testing (NDT)	10	7.6	40	28.8
Over pressure test	2	1.5	—	—
Catastrophic failures				
Associated piping	—	—	4	2.8
Operational reason	4	3.0	9	6.5
Vessel quality	3	2.3	3	2.1
Totals	132	100.0	139	100.0

[a] Based on study by Phillips and Warwick [5], 1962-1967, and Smith and Warwick [6], 1967-1972.

Relevant Experience of Nuclear Pressure Vessel Integrity

O'Neil and Jordan [3] estimated that there was approaching 2000 reactor vessel years of experience including all known military reactors. This figure is consistent with the subsequent ACRS study [4]. No category A or B failures have been reported. However, this experience is insufficient to meet the targets derived in Table I, and is a situation which will not be corrected with time since it is unlikely that much more than 10^4 vessel years of experience will be accumulated before the end of this century.

Experience and Estimates Based on Nonnuclear Pressure Vessels

United Kingdom

A survey [5] was carried out by Phillips and Warwick covering some 10^5 vessel years of experience in 12,700 class I vessels over the period 1962-1967. This was extended by Smith and Warwick [6] who covered a similar number of vessel years over the period 1967-1972. The results of these studies are summarized in Table II, from which it would appear that

TABLE III

Distribution of Causes of Failure (or Withdrawal from Service) Found in UK Pressure Vessel Survey

	1962-1966		1967-1972	
Cause of failure	Number of cases	%	Number of cases	%
Cracks	118	89.3	117	84.2
Corrosion (including stress corrosion, assisted fatigue, and wastage)	2	1.5	—	—
Maloperation (including water shortage and control malfunction)	8	6.1	10	7.2
Material or fabrication causes	3	2.3	11	7.9
Creep	1	0.8	1	0.7
Totals	132	100.0	139	100.0

catastrophic failure due to loss of vessel integrity occurred in some 2% of the service failures reported, correspondent to a failure rate of about 3×10^{-5} per vessel year. However, a more detailed study of the failures reported above shows that no vessel failed catastrophically in the population under review, the reports more usually referring to piping or attachments, small components, or nonrelevant operating conditions. Zero failure reports would imply a failure rate of 1.5×10^{-5} at the 95% confidence level or 2.3×10^{-5} at the 99% confidence level, assuming failures to be normally distributed. Studies by the ACRS [4] reach a similar conclusion on the relevance of the UK failures quoted.

If the means of identifying these service failures are considered, an interesting pattern emerges. From Table II it can be seen that something like a third of the defects indicated their presence by leakage from a noncritical defect. If such experience could be transferred to the nuclear pressure vessel, then it would appear that the target for category C can be achieved, i.e., 38 cases in 10^5 vessel years, or about 4×10^{-4} per vessel year.

If the other "potentially dangerous failures" are examined (this is the survey terminology equating to category A and category B "critical defects" in this section), it will be seen that in the first survey, visual examination revealed 57% of the defects and dominated NDT which only revealed 7.5%. The situation was reversed in the second survey where visual

7. The Integrity of Pressure Vessels

TABLE IV

Distribution of Causes of Cracks Reported in UK Pressure Vessel Survey

Cause of crack	1962-1967			1967-1972		
	Number of cases	% of cracks	% of all failures	Number of cases	% of cracks	% of all failures
Fatigue (including mechanical and thermal fatigue)	47	40	35.6	20	17.0	14.5
Corrosion (including stress corrosion, assisted fatigue, and wastage)	24	20.3	18.2	4	3.4	2.9
Preexisting from manufacture	10	8.4	7.6	48	41.0	34.6
Not ascertained	35	29.6	26.5	18	15.4	12.9
Miscellaneous	2	1.7	1.5	27	23.2	19.3
Totals	118	100.0	89.4	117	100.0	84.2

examination revealed 24.5% of the defects, while NDT revealed 29%. This is not thought to be due to anything more significant than the widespread use of NDT, either independently or as a means of backing up a visual examination. In neither survey was pressure testing shown to be of any great significance (1.5% and 0%, respectively).

The surveys also examined the physical reasons for failure or withdrawal from service and these are summarized in Table III. In both surveys it was shown that cracks dominated. Material or fabrication causes are not shown to be very significant in the first survey (2.3%), and while significantly more so in the second (7.9%), they still represent a small fraction of total failures. It is difficult to draw any firm conclusion from the figures, but it seems probable that they may only represent an improvement in the postfailure investigation practices. It is of interest to examine the principal causes of cracks which are summarized in Table IV.

This appears to show a disturbing trend in the presence of cracks preexisting from manufacture. They have risen from 8.4% of the events to 41%. It is difficult to quantify how much of this increase is a function of improved postincident investigation, but it is thought that it is unlikely that it

TABLE V

Summary of In-Service Experience of Nonnuclear Pressure
Vessels Considered Relevant to Nuclear Plant

Item	Source	Number of vessels	Total vessel years	Number of disruptive events	Disruptive failure frequency per vessel per year (upper 95% confidence level)
1	Phillips and Warwick [3]	$\sim 2 \times 10^4$	10^5	0	3×10^{-5}
2	Smith and Warwick [6]	$\sim 2 \times 10^4$	10^5	0	3×10^{-5}
3	Kellermann et al. [7,8]	7×10^3	6.7×10^4	0	4.5×10^{-5}
4	Kellermann and Siepel [9]	2.4×10^5	3.77×10^6	5^a (assumed)	0.28×10^{-5}
5	EEI-TVA [7]	1×10^3	1×10^4	0	30×10^{-5}
6	EEI [8,9]	5×10^3	2.27×10^4	0	13.2×10^{-5}
7	ABMA [8,9]	6.8×10^4	72.3×10^4	0	0.42×10^{-5}
8	Combined data		4.79×10^6	5	$2.2 \times 10^{-6 b}$

^{a}Estimate based on comment that the frequency of events leading to "severe vessel damage" was 3 in 10^6 yr. This corresponds to 11 disruptive failures in the sample. However, it was reported that 57% of these failures arose from maloperation having no counterpart in the nuclear field, leaving a total of 5 disruptive events to be considered.
^{b}This is equivalent to a mean disruptive failure rate of about 10^{-6}/vessel yr.

is due to a real reduction in manufacturing standards or quality control. Fatigue and corrosion on the other hand have dropped over the same period from 60% to 20%. While this could well be indicative of an improvement in understanding of mechanisms of fatigue and corrosion, the size of the improvement seems greater than might reasonably have been expected over such a short period. On the basis of available evidence, it is doubtful if any conclusion can be drawn other than that preexisting cracks and fatigue (including corrosion assisted fatigue) are the principal causes of pressure vessel failure.

German Experience

The German Institut für Reaktorsicherheit der Technischen Überwachungs—Vereine e.V (TUV) has a statutory responsibility for surveying pressure vessels and collating failure information. This has been collated and published in a series of papers [7,8] by Kellermann and his co-workers. Two principal

7. The Integrity of Pressure Vessels

studies have been made covering 67,000 pressure vessel years and 3.7×10^6 pressure vessel years, respectively, and are summarized in Table V. The reports are not so specific in the breakdown of the categories or causes of failure, but in the major survey it is stated that "severe pressure vessel damage" is about 3×10^{-6} per vessel year. In a population of 3.8×10^6 pressure vessel years, this corresponds to a total of 11 disruptive events. However, Kellermann et al. state that 56% of the failures were "mostly caused by failures in operation." Examination of the data indicates that only 5 of these failures are relevant in the analysis of failure of nuclear pressure vessels. The 95% confidence level is not particularly sensitive to this assumption. The following figures illustrate this point.

Assumed number of disruptive events	Upper 95% confidence failure frequency for item 4	Resultant "combined data" result (Table V, item 8)
0	0.8×10^{-6}	6.36×10^{-7}
5	2.8×10^{-6}	2.2×10^{-6}
11	4.9×10^{-6}	3.84×10^{-6}

US Experience

Relevant service experience with pressure vessels in the US has been summarized in the ACRS Report [4]. Three separate studies were reported. The Edison Electric Institute—Tennessee Valley Authority (EEI-TVA) analyzed data supplied by 55 utilities on boiler steam drums fabricated to section 1 of the ASME code. A total of 1033 vessels were examined having a cumulative in-service experience of 10 vessel years of operation. No disruptive failures were reported, but in 10^4 instances subcritical flaw growth was detected. As with the Phillips and Warwick survey, most of the cracks were associated with branch pipe attachments and most (if not all) were caused by fatigue. The second survey was carried out by the Edison Electric Institute (EEI) for the USAEC. Again, the source of the experience was from steam drums and pressure vessels in a fossil-fired power plants, but the population of vessels was restricted to those vessels (a) built between 1959 and 1968 to the ASME Code (section I or VIII), (b) with a wall thickness greater than 1.5 in., and (c) with operating conditions or pressure greater than 600 psig and temperature between 250 and 800°F.

Experience with a total of about 5000 vessels was reviewed covering 22,000 vessel service years. No disruptive failure and only 1 noncritical failure were reported. This study is very comparable in its scope and findings to that undertaken by Kellermann et al.

The third study was based on the experience of the American Boiler Manufacturers Associated (ABMA). The survey covers an estimated number of 68,000 vessels manufactured to ASME Code (sections I and VIII) requirements since 1939 and corresponding to about 723,000 vessels service years. Once again, no disruptive failure was reported. The incidence of nondisruptive failures was not given.

Summary of Relevant Nonnuclear Experience

The disruptive failures are rather more difficult to analyze. Since no relevant failures are reported in the smaller samples, the 95% upper confidence failure rate quoted is merely a function of the assumption of normal distribution and the sample size. Only the major Kellermann and Siepel [9] study quotes any actual failures (equivalent to about 10^{-6}/vessel yr). If all the studies in Table V are considered valid, then their statistical value is improved if they are added together. The only argument against such a technique is the possibility that the same events may be quoted twice in the US or German studies. However, in each case the major study is so much larger than the others that double counting is not likely to have any statistical significance. Thus, if all the results are summed, the final disruptive failure rate becomes 2.2×10^{-6}/vessel yr at the upper 95% confidence limit. This is not inconsistent with the conclusions of Ref.[2] that disruptive failure rate of relevant conventional pressure vessels is less than 10^{-5}/vessel yr.

Extrapolation to Nuclear Vessels

The category A target failure rate for nuclear vessels, as summarized in Table I, is 10^{-6}/yr. While it is possible that vessels currently being used in nuclear installations meet this target, this cannot be demonstrated by reference to experience in the nuclear field, or by direct evidence from the nonnuclear field as summarized in Table V. This is because of the different material, construction, nuclear background, etc., and raises two questions:

(a) Can reactor pressure vessels generally be made to the required standard?

(b) Is the specific vessel up to the required standard?

7. The Integrity of Pressure Vessels

In reliability engineering there are well-established techniques for synthesizing overall failure rates by consideration of the component parts which contribute to this overall failure rate. Such consideration may be based on direct statistical evidence or, where this is not available, on informed judgment. Such methods are invaluable in showing up the degree of reliance on people or processes, and assist in deciding whether or not such reliance is soundly placed. These techniques are more commonly applied in the field of electronics or in mechanical systems using mass production components, but in this section consideration is given to the use of such methods in the production of pressure vessels.

For a pressure vessel, the probability of gross vessel rupture occurring between service inspections might be expressed as the likelihood of failing to meet the requirements of good design, material, and construction standards, as currently practiced, modified by the combined failure probability of all testing techniques applied between service inspections. These tests are some or all of the following:

a pressure test,
stress wave acoustic emission test,
ultrasonic test,
leakage prior to gross failure, and
visual examination.

Failure in Design, Materials, and Construction

The probability of a failure in design would require a gross error in estimated duties, e.g., stressing, fracture mechanics, critical crack size, and crack growth. There are no statistical data on the frequency of such errors, and there are considerable uncertainties in all of these headings. However, pressure vessel designs are not carried out *ab initio*, but rather as a series of limited extrapolation from previous platforms on the basis of established design codes, where relevant. Where a number of vessels with novel features are built in the same class or series, it may be necessary to take special measures to clear the design for the first of class only. The reason for this need is that while some design deficiencies can be revealed by test and inspection procedures before service, others will not be revealed until the vessel has been in operation for a significant number of service cycles. This means that a novel vessel design is particularly at risk until it receives the first in-service inspection. Thereafter the risk is mainly due to increasing cycles and deteriorating properties, both of which should have been predicted and can be monitored.

The probability that the vessel will not have material properties up to the standard specified by the design is significant, and this is particularly true where alloy steels and other special materials are involved. Such mistakes, once made, may be difficult to detect in initial inspection unless they result in some identifiable defect, and may produce defects which grow at a completely unacceptable rate resulting in gross failure before the first in-service inspection. Defects which are identified after a period of service can be analyzed before further service is permitted, but to ensure that the rate of gross vessel failure due to material deficiencies does not become excessive requires the most rigorous quality control and oversight. This will require material identification, bonding, and inspection to standards not commonly achieved in the pressure vessel industry.

The probability that there will be some errors in construction that could lead to gross failure due to, say, faults in heat treatment, fabrication, repair, or after a process inspection should be low but cannot be ignored. The overall failure rate of a vessel suffering from some failure in design, materials, or construction might appear to be greatest during the period up to the first inspection.

Failure of Pressure Test to Reveal Potential Failure

A successful over-pressure test carried out on a pressure vessel will indicate a high probability of freedom from a critical defect—but not universally so. If, subsequently, the vessel is operated at some lower pressure, it may be possible with knowledge of the crack growth mechanism to estimate the number of cycles that may be permitted until the next pressure test. It is important to recognize that the pressure test applies a membrane load, although in service the vessel may be subject to thermal shock, fatigue, bending loads, etc. Furthermore, although a pressure test may, by causing failure, demonstrate gross deficiencies in design, material properties, or construction, these are more likely to require a number of cycles to propagate defects to a dangerous extent. Even in the well-documented Cockenzie vessel, failure did not occur until after four over-pressure tests. The failure was due to a defect that could have been readily detected by recognized nondestructive testing methods.

The limited evidence reproduced in Table II shows that over a five-year period only two potentially serious defects (1.5% of the total) were detected.

It appears that pressure testing has served an important part in the growth of pressure vessel technology, but no longer dominates the inspection regime unless coupled with other forms of NDT such as stress wave acoustic testing—a technique that has yet to be fully established.

Failure of Stress Wave Acoustic Test to Reveal Potential Failure

When a vessel is loaded to the point where defects are propagating, even at the crystal level, bursts of elastic energy waves known as stress wave emissions (SWE) take place. By the use of suitably placed sensors, these bursts can be detected and located. The system has potential for initial over-pressure testing, periodic over-pressure testing, or on-line monitoring, although these have yet to be convincingly demonstrated.

The likelihood that suitable equipment, once commissioned, will fail to detect a signal in a sample situation is small; but the interpretation of the signal to decide whether it is important or not depends on the physical nature of the defect and a detailed stress and fracture mechanics analysis. The chance of successful identification of an important defect in a complex structure is still very uncertain.

On repeat pressure tests the above critique is also likely to be true, except that the emissions being analyzed are likely to be smaller until the pressure of the initial test is approached, particularly when the tests are close together in time ("Kaiser effect"). Moreover, the vessel will be less conveniently placed for testing and more extraneous signals may be introduced.

Failure of Ultrasonics to Reveal Potential Failure

Ultrasonic examination has now been developed to the stage where defects which could be of safety significance (say, about 1 cm) are within the detection capability of the inspection equipment as defined by calibration blocks. A full baseline examination of the type defined in the ASME code requires detailed examination of any area considered critical on the basis of a typical stress analysis and fracture mechanics evaluation. The problem becomes one of reliability of the process. Subsequent examination is on a sample basis over a typical 10-yr cycle. The process relies mainly on the skill and responsibility of the operator and the integrity of his equipment, and in some circumstances may not be capable of detecting defects in uniquely unpreferred orientations.

This inspection is taking place on a vessel where process inspection might have been expected to locate any obvious defects. Furthermore, in the case of an initial inspection, any potential defects that are initiated and propagated by fatigue will not yet be present. The probability that ultrasonic testing will fail to detect a potentially serious defect is assessed at $1\text{-}10^{-1}$ per test over the first inspection period and $10^{-1}\text{-}10^{-2}$ over subsequent inspection periods for defects up to about 2 1/2 cm. For maximum effectiveness the baseline ultrasonic test should be carried out before as well as after the final commissioning pressure test.

Failure of Visual Examination to Reveal Potential Failure

Table II, prepared from the survey of Refs.[5, 6], suggests that visual examination has been the most useful method of detecting pressure vessel failure, and ASME XI attaches considerable importance to surface inspection. It may be, however, that the table only indicates the present strong bias to visual examination in routine in-service inspection of conventional vessels by insurance inspectors.

In a nuclear pressure vessel, visual in-service inspection must be carried out remotely, either by optical means or by television. Enormous improvements have been made in the quality of television inspection, which can also detect under water virtually anything that can be detected by direct surface examination. However, nuclear vessels are likely to be stainless-steel clad on the inside surface, and it is possible that a potentially critical defect which propagates into the parent metal from the cladding interface would not be seen on examination of the exposed clad surface. However, if the clad is intact, the propagation of cracks in the parent metal is not enhanced by the environment and is at a lower and much safer rate. In-service inspection from the outside of the vessel may be limited by design consideration.

Visual examination is an ideal method of detecting physical damage to the reactor structure and core internals but may not be highly reliable for detecting potentially dangerous pressure vessel defects, and the probability that it will fail to locate such flaws is tentatively assessed at $1\text{-}10^{-1}$ per inspection.

Failure of Leakage to Reveal Potential Failure

In conventional vessels, as indicated in Table II, leakage provides a useful advance warning of a potentially dangerous situation. However, the conventional vessels on which the study was based tend to be comparatively thin walled (less than 2 in.) and made of ductile materials. In nuclear vessels, wall thicknesses of 4-12 in. are normal, and alloy steels are used.

7. The Integrity of Pressure Vessels

As a result, there are areas where the critical defect size is considerably less than the through thickness of the wall, and the classic "leak before break" situation may not be applicable. Analyses based on tenuous calculations are made, but little convincing experimental evidence is available,* and the probability that leakage will fail to reveal a potentially catastrophic failure of the pressure vessel is tentatively assessed at $1-10^{-1}$ over an inspection cycle.

Discussion

Many inspecting organizations, using sophisticated techniques, might expect to fare better against defects likely to lead to vessel failure than these assessments suggest. This may well be true for deficiencies in construction, and to some lesser extent for material deficiencies, but is a difficult argument to sustain in the case of design deficiencies which may not produce any physical manifestations until after a period of service. Indeed it is possible to postulate circumstances that would completely negate all the inspection processes currently available (on-line acoustic testing not yet being available). It is for this reason that a complete and independent assessment of the design is a necessary part of the overall quality control of a nuclear pressure vessel.

Failure in vessels of the type under discussion are likely to be few, and direct evidence on the reliability of in-service inspection techniques is limited. However, as indicated in Table V, the mean catastrophic failure rate for conventional pressure vessels based on world data is about 10^{-6}/yr. If a normal distribution is assumed, the 95% upper confidence level does not exceed 2.2×10^{-6}/yr. No convincing arguments have been advanced to show that nuclear vessels cannot be built to at least comparable standards. Inspection should confirm that such standards have been met, and are maintained throughout service life. Confidence that inspection standards can currently be met is difficult to demonstrate. Evaluation programs are being undertaken in the UK and overseas, but fully convincing results are unlikely to be available for some years. Meanwhile, limits on engineering capability are likely to restrict equipment reliability to a "fail-danger" rate of about $10^{-2}-10^{-3}$ per inspection cycle, and operator performance to about 10^{-1} perhaps rising to 10^{-2} per inspection cycle, resulting in an overall "fail-danger rate" of about 10^{-1} per inspection program.

*The No. 5 US HSST vessel is of interest in this respect.

The Probability of Failure of Nuclear Pressure Vessels

The foregoing discussion would lead us to believe that the small, residual possibility of pressure vessel failure through error in design, or material, will be significantly reduced by the various tests and inspections which are now applied. It may be difficult to establish evidence that these techniques can be applied with a 99% chance of success, which would be needed to effect a 100-fold reduction in the already small, residual risk. However, it would be difficult to argue that they would not achieve a 90% success rate or a 10-fold reduction in risk. If applied to conventional vessel experience, the failure rate would be reduced to 10^{-7}/vessel yr or 2.2×10^{-7} at the 95% confidence level. This is a rate not inconsistent with current studies based on probabilistic fracture mechanics. However, such studies are still at a very early stage in development and beyond the scope of this work.

Conclusions

1. Steel pressure vessels for nuclear reactors must be designed and produced to the highest standards currently available, but even these cannot be demonstrated to be to the standard required. Further improvements in confidence can only come from additional independent detailed assessment of design and an independent regime of inspection and testing.
2. Quantitative safety and reliability goals for such inspection and testing can be specified and interpreted in terms of acceptable failure rates for pressure vessels. Some tentative values were proposed in this chapter.
3. Periodic inspection is fundamental to the process of structural validation, and despite the difficulties associated with radioactive areas, inspection must be extended to such areas by the development of sophisticated equipment for remote handling and data presentation.
4. More attention needs to be given to the avoidance of nugatory inspection. The latest code practices endeavor to ensure that the areas selected for safety inspection are apposite. However, the effectiveness of inspection can only be measured by the degree of success in its application to "significant" defects. Some tentative values are proposed which are consistent with current experience.
5. Using reasonably cautious values for the overall capacity of additional in-service inspection, and making limited demand on the capability of containment and other engineered safeguards to reduce the consequence of catastrophic

7. The Integrity of Pressure Vessels

vessel failure, the safety standards derived from the risk line in Fig.1 appear to be met if nuclear vessels meet the standards apparently being achieved by conventional pressure vessels and are subjected to further inspection to further reduce the probability of defective vessels remaining in service.

References

1. Beattie, J. R., Bell, G. D., and Edwards, J. F. Methods for the Evaluation of Risk, AHSB(S)R159 (1969).
2. USAEC. Safety of Nuclear Power Reactors (Light-Water Cooled) and Related Facilities (July 1973).
3. O'Neil, R., and Jordan, G. M. Safety and reliability requirements for periodic inspection of pressure vessels in the nuclear industry, presented to the Institution of Mechanical Engineers, 1972, Paper C48/72.
4. ACRS. Integrity of Reactor Vessels for Light-Water Power Reactors, ACRS Rep. WASH-1285 (1974).
5. Phillips, C. A. G., and Warwick, R. G. A Survey of Defects in Pressure Vessels Built to High Standards of Construction and Its Relevance to Nuclear Primary Circuit Envelopes, AHSB(S)R162 (1968).
6. Smith, T. A., and Warwick, R. G. The Second Survey of Defects in Pressure Vessels Built to High Standards of Construction and Its Relevance to Nuclear Primary Circuits, UKAEA, SRDR30 (1974).
7. ACRS. Integrity of Reactor Vessels for Light-Water Power Reactors, WASH-1285, p. 67 (1974).
8. Kellermann, O., Kraegeloh, E., Kussmaul, K., and Sturm, D. Considerations about the Reliability of Nuclear Pressure Vessels—Status and Research Planning, Paper 1-2, Pressure Vessel Technology, Part 1, Design and Analysis, 2nd International Conference for Pressure Vessel Technology, San Antonio, Texas (1973).
9. Kellermann, O., and Siepel, H. G. Analysis of the Improvement in Safety Obtained by a Containment and by Other Safety Devices for Water Cooled Reactors, presented at IAEA Symposium on the Containment and Siting of Nuclear Powered Reactors, Vienna, 1967, Paper SM-89/8.

8
Thermal Reactor Safety
J. H. Bowen

All the reactor systems in use for power generation—thermal and fast—have stable power characteristics, at least to first order. If the control rods or the coolant flowrate are changed so as to increase power, the fuel temperature will rise and oppose further increase in power—the change is self limiting. Controllability of the nuclear reaction is not a difficult problem. Safety consists in matching the heat produced in the fuel to the heat removed by the coolant. A mismatch results in a different level of core temperature which, if allowed to persist, may cause failure or melting of the core materials.

As soon as a mismatch occurs and is picked up by the sensors, the power generated may be suppressed by scramming the control rods very quickly (Fig.1). As far as possible a fast scram is not used—to avoid the loss of station power and to avoid the over cooling of engineering structures.

However, in most cases, heat production and heat removal can be easily matched, with the particular exception that heat production cannot be reduced to zero; "afterheat" persists due to the alpha, beta, and gamma decays from previously fissioned atoms and other materials following neutron capture. One minute after the neutron reaction has stopped, the afterheat amounts to 3 1/2% of the previous steady power level and drops slowly to 0.6% after one day (Fig.1 of Chapter 6). The removal of this heat is an important aspect of reactor safety.

Another related aspect is the thermal inertia of the fuel. When the neutron reaction is stopped, the fuel is hot and normally has a temperature distribution (see, for instance,

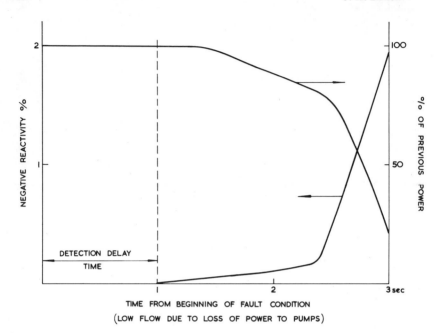

FIG.1 Typical PWR illustrating quenching of nuclear reaction by shutdown (SCRAM).

Fig.2). If heat production and removal cease simultaneously, the hotter parts of the fuel cool and the cooler parts heat up, both reaching, theoretically, an equalization temperature. One convenient measure of the heat content of the fuel is the number of seconds of operation of the reactor at full power which would heat the fuel to this temperature.

The idea of sudden insulation of a pin is an extreme picture, but one which approaches realization in "stagnation" or "vapor blanketing" either alone or in combination. Nevertheless, a break in a simple high-pressure system in which either liquid or gas is circulating can result in a drastic reduction of coolant flow, giving a condition approaching the sudden insulation postulated above. If the flowrate of a liquid coolant is reduced sufficiently, it will begin to boil at the outlet of the fuel channels. The vapor requires a larger driving pressure than the liquid (for a given coolant mass flowrate); this causes a further reduction of flow, so the process has a tendency toward instability. The flow regime could change from a liquid coolant to a vapor-filled channel in a time comparable with the transit time of coolant from inlet to outlet of a channel—probably a fraction of a second. Finally, either alone or in combination with the foregoing, there is the possibility of "dry-out." If the heat flux is progressively

8. Thermal Reactor Safety

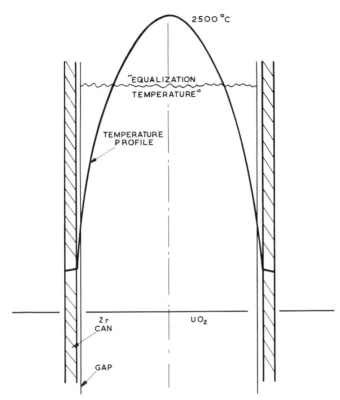

FIG.2 Temperature distribution across peak PWR fuel pin in normal running (approximate).

increased, coolant may commence to boil at the can surface although most of the fluid in the channel is below boiling temperature (subcooled nucleate boiling). This is a powerful heat transfer process; but at a much higher heat flux again, the vapor bubbles tend to coalesce and vapor-blanket (i.e., partially insulate) the pin. The equalization temperature is often used in getting a rough estimate of the temperature caused by such effects. We consider how this condition is detected; how many seconds are required, therefore, to get the scram rods moving; convert these seconds of power to fuel pin temperature increment; and add this to the equalization temperature.

The heat capacity of the fuel pin may be helpful if the power being generated in the fuel is rapidly increased. The nuclear power may considerably increase (say, even double), but the power to the coolant is cushioned by the storage of heat in the fuel. For example, in order to transfer heat to the coolant at twice the normal rate, approximately twice the normal full-power seconds must be stored in the fuel pin. Thus, steam blanketing is delayed, giving the opportunity to scram the reactor on signals from high-neutron power.

For "normal" control rod withdrawal rates (say, 1% reactivity per minute), reactor power, heat flux, and Doppler feedback are all essentially in equilibrium and it is not necessary to invoke this argument—power overshoots are minimal.

In all the current practical power reactors, coolant is pumped through the reactor core which has been designed for the twin requirements of neutron economy and heat transfer. A local change to the design geometry might have little effect on "reactivity" (i.e., neutron economy) but could have a large local effect on heat transfer. An example would be the blocking of a flow path, perhaps by a foreign object, or perhaps by accidental breakage of fuel or structure. Coolant flowrate and coolant and fuel temperatures are usually measured, but it is not practicable by this means to claim the ability to detect very local overheating. Usually the most definite sign would be a rise in fission product activity, indicating that local overheating had locally damaged fuel cladding. Some such small increase in radioactivity in the circuit is allowed for in the containment margins; in this introduction, we wish to consider whether the local overheating has more serious implications. Of course, if the original blockage has caused some fuel element failures, the debris from these may cause further blockage. This could be countered by shutting the reactor down within a suitably short time from the original high-activity signal.

The chemical effects of this local overheating must be considered. In the British Magnox reactors, it was recognized as a possibility that the magnesium canning would burn at around its melting temperature in CO_2 and thus might cause a fire in one channel. The calculations aimed at discovering whether or not the fire could spread to other channels; it was concluded that it could not and in St. Laurent des Eaux, Chapelcross 1, and Latina, only the one channel was involved with no serious effect on safety or reactor utilization. With Zircalloy cladding in a water reactor, there is an exothermic reaction but there is no flame, as with magnesium. The metal/water reaction heat is added to the neutron power and the afterheat; it is about equal to the latter at 1300°C and is dominant at, say, 1400°C, but, of course, disappears again as the Zr is consumed. The aim would be, taking account of all heat sources, to show that the reactor has been scrammed while the blockage and overheating are still very local with respect to the whole core. While there should be little doubt about this, even allowing for the slowness of activity signals and manual operation of scram, the melting of that region originally affected

8. Thermal Reactor Safety

by the blockage remains a possibility. In a sodium-cooled reactor the further possibility that this molten fuel material may "interact thermally" (as opposed to chemically) with the coolant and produce a local pressure by suddenly producing sodium vapor is taken very seriously. As regards water cooling, several differences from sodium may be taken into account. The ratio of vapor volume to liquid volume is very different: 1000:1 for the atmospheric pressure sodium coolant; 5:1 for a BWR at 60-atm pressure. The implication is that the work available in expanding from one state to the other is similarly smaller. Nevertheless, the possible mechanical work could still be considerable. The sudden development of this mechanical work is only possible if heat is transferred from the source (the fuel) to the working substance (the water) in a few milliseconds, and this requires intimate mixing of the two equivalent to millimeter size agglomerates. In some theories of the dispersion process, this is physically impossible at the LWR pressure, but not everyone would accept this. On the whole, the safety argument rests upon the greater ability of an LWR core to accept the disruptive event. In the fast reactor there is the possibility of jamming of control rods. Not only does this seem less likely in the larger LWR cores, but we may also reason that any core rearrangement in a thermal reactor which is sufficient to jam all control rods is likely to shut the reactor down by changing the core shape, lattice pitch, and properties of fuel to moderator—whereas in the fast system, the safeguard of the moderator (i.e., that only a precisely defined arrangement is critical) no longer applies.

In summary, therefore, the consequence of a local melt out in any of the thermal reactors should only be the release of fission products to the circuit from the fuel immediately affected. The normal fission product monitoring devices are designed to detect the early onset of this (e.g., say, a square centimeter of fuel uncovered), and having shut down, the damage should not progress further.

Fission Product Releases

In the following sections, the factors which influence the course of events under three general headings in which reactor accidents can be envisaged—reactivity faults, loss-of-flow faults, and loss of pressure—are discussed. For present purposes we may classify the outcomes as follows:

(i) Normal operating limits hardly exceeded; virtually no fission product release within the circuit; still less to atmosphere.

(ii) A few percent of the cans fail; these release perhaps 5% of their iodine and gas inventory; some circuit cleanup may be necessary but release to atmosphere of order 1 Ci ^{131}I.

(iii) General melt out of fuel; most of the iodine, cesium, and gases released from this fuel.

(iv) Some fuel is vaporized. This is only possible in thermal reactors if a very fast reactivity increment is possible. In the discussion below, this arises only in the context of rapid control rod ejection in an LWR, and in that case is prohibited by operating limits.

In the following discussion, use is often made of condition (iii) to define the reliability sought from the combined safeguards and the accident frequency; as described in Chapter 2, for a large reaction, this "potential" release is about 10^7 Ci ^{131}I with associated fission products. The rod ejection possibility is hardly capable of volatilizing the fuel extensively as it is localized by a flux spike which develops in the empty control rod position, so the volatilization of a few percent of the core has been taken as still broadly equivalent to the 10^7 Ci release.

Reactivity Faults

Although a thermal reactor lattice is close to optimum as regards neutron economy, there are still several ways of increasing reactivity. The first is provided by the control rods themselves. Before and during start-up the rods are fully or partly inserted and if withdrawn too quickly may cause a mismatch between cooling and power generation. Even when the reactor is running steadily on full power, some control rods remain partly inserted, ready to compensate for fission product poisoning and depletion of the fissile inventory of the core. The larger the reserve of reactivity held in this way, the longer the time that the core will operate before refueling. On the other hand, this is provided at the expense of higher fuel initial enrichment; this economic optimization generally results in 3% or 4% of reactivity available in control absorber throughout most of the core life. One general way in which this could be accidentally added to would be if the mechanism for withdrawing the rods were accidentally actuated. For a PWR or a gas-cooled reactor, we obtain the consequent rate-of-rise of core temperature by dividing the rate of addition of reactivity* (a design figure limited by

*"Reactivity" is a measure of the increase in the neutron population of one generation over that of the preceding generation.

8. Thermal Reactor Safety

the speed and gearing of the control rod motors) by the temperature coefficient—about $10^{-4}/°C$ for both reactor systems. Thus, an addition rate of 0.1%/min causes a 10°C/min temperature rise of the core. The reactor power adjusts itself to achieve this. For the PWR, the first danger point would probably be in the coolant pressure which would rise to safety valve setting in 2-3 min. In the meanwhile, many alarms should sound, and the reactor should trip—from temperature, neutron flux, pressure, and perhaps other signals. A BWR is a little different, being a constant-pressure system. The boiling rate would increase at essentially constant temperature. The power/reactivity coefficient (about -4.10^{-4}% increase in power) may be used to estimate the rate of increase. Neutron power and steam pressure rise would be available as trip parameters.

For gas-cooled reactors, it is possible for the moderator temperature coefficient to be influenced by a fission resonance in ^{239}Pu; then, on a time scale of minutes, the reactor could have a tendency toward instability, i.e., as the moderator temperature increases, reactivity is increased, leading to further temperature increases. The reentrant cooling on AGRs, in which coolant at inlet temperature cools the moderator, before continuing its flow path to cool the fuel, strongly stabilizes moderator temperature, so that the reactivity change occurs so slowly that it is easily countered by trimming control rods.

We now consider the implications of faster rates of reactivity addition without, for the moment, specifying how this might happen. If the rate were increased tenfold to 1%/min, this would present no challenge to the scram system which can operate in about 1 sec. It would allow less time for operator intervention, and thus increase the required reliability on the scram system, but multiple signals would still be available. A significant change occurs only when the rate of addition of reactivity is so fast that the temperature and power coefficients diminish in value, since the heat then does not have time to spread among the core components—in particular, to heat the moderator. If the heat is essentially confined to the fuel due to the rapidity of the transient, the reactivity coefficient is of order $10^{-5}/°C$ in all cases. A reactivity addition of the order of 1% can then add 1000°C to the peak fuel temperature, normally for a PWR 2600°C, thereby melting some UO_2 at 2800°C or vaporizing some at 3400°C.

The generation times themselves are variables, depending mainly on delayed neutron lifetimes. Thus, reactivity and generation times enable the reactor physicist to calculate the rate at which the neutron population, and hence the power, will change. Reactivity is simply a positive or negative number, e.g., 0.1%.

Considering now the practical ways in which reactivity addition might be imagined to occur, the first is by control rod removal. The gas-cooled reactors, with their large cores and large numbers (say, 100) of control rods, are inherently not at risk from the removal of a single rod, worth perhaps 0.1% reactivity. A rod in an LWR system is capable of being worth considerably more—say, 2%, due to the smaller neutron migration area, but operational measures are taken to limit the worth of the rods.

On PWRs, the current practice is to effect long-term reactivity control by a neutron absorbing boron salt dissolved in the moderator; it is added slowly as a weak solution at the pumps which thus mix it with the moderator, and then removed by bypassing part of the coolant stream through ion-exchange beds. The role of the control rods becomes minimal for short-term adjustments; the pattern of the rods and their depth of insertion may be easily controlled so that no rod would be "worth" a dollar of reactivity, that which would just cause prompt criticality if ejected. Below prompt critical, approximate calculations establish reactivity limits at which fuel will not melt. These operational precautions make it far less critical as to whether a given rod may be inadvertently removed or ejected.

Some LWRs have hydraulic actuation of control rods; obviously fail-danger fault conditions must be prevented by design. If a control rod pressure housing failed, a rod could be ejected by coolant pressure; a secondary hold down is usually provided against this. On a BWR, the rods are inserted into the core from below; to start the reactor they are pulled down. The possibility has been recognized that a control rod (the "blade" section) could be detached from its "pusher" which might be withdrawn leaving the blade behind. Subsequently this could fall out of the core. The design ensures that the resistance of the water limits the velocity of the blade in this fault condition to a safe value of about 5 ft/sec. The pattern of control rod withdrawal is also designed so that no one rod should release too much reactivity if completely withdrawn at any point in the cycle; the increment is thus limited to about 1%.

Although this exceeds the value of reactivity which would be prompt critical (the BWR, for reasons of water chemistry, does not utilize the uniform liquid boron poison), it is possible to show by more sophisticated calculation that this reactivity release, at the limited rate, does not cause fuel pins to fail catastrophically.

During core refueling, control rods and fuel elements are moved at rates which are not so well-defined as in normal operation. Of course, the intention is that the reactor should be

8. Thermal Reactor Safety

safely subcritical at this time. There is no doubt from the neutron flux measurements about whether it is subcritical, but the evidence as to how much it is subcritical, whether 5% or 15%, is rather indirect. The reactor core is open to the containment at this time, and the fission product emission to atmosphere from damaged fuel could be serious. The best safeguard seems to be a refueling condition in which the reactor is subcritical by a large margin (even 20%). It seems that this can be achieved for a PWR by liquid boron poison, but is not so practicable for a BWR for reasons of chemistry, and closer control is therefore necessary.

Reactivity could be added to an LWR by cooling the moderator. As previously explained, only a very fast reactivity addition rate represents a danger and the only possibility would seem to be if cold water could enter the core more or less as a "plug flow." Taking $10^{-4}/°C$ for the temperature-reactivity coefficient, we see that the full flow from a circuit of a 3-circuit PWR, entering at, say, 100°C (the other circuit being at 300°C) could just about produce the sudden drop of 100°C required in the average core temperature to give the major danger. This accumulation of a large slug of cold water is designed out by avoiding fast-operating valves between the separate circuits, and by keeping a hot water bleed through all circuits at all times.

In a BWR, where some 10% of the core volume is occupied by steam, the steam is much less effective than water as a moderator, so if water replaces some steam, reactivity is added. The void coefficient of a BWR is typically 10^{-5}% of void. This process of replacing steam by water occurs if resistance is introduced against the flow of steam to the turbine, thus increasing the reactor pressure. In particular, if the turbine stop valve is closed, reactor power tends to increase at a time when it should be reduced, thus further increasing coolant pressure. Although a turbine bypass is usually provided to cushion this process (the safety of the turbine may require a rapid closure of the stop valve), safety really rests upon scramming the reactor. As described earlier, multiple signals exist—neutron power, coolant pressure, and conditions at the turbine.

The SGHWR should not have the characteristic just described—its moderation should depend little on the light water in the pressure tubes—but there could be a possibility of overpressurizing the calandria if a pressure tube burst into the calandria space, leading to expansion or bursting of the outer shell of the calandria (reactivity removal) and the hoop compression of the calandria tubes (reactivity addition). The in-core circuit and calandria are, therefore, engineered to a standard of reliability which reduces the possibility of such a reactivity accident to an acceptably low level.

The AGR is not prone to any of the reactivity incidents just described, all of which have the character of "initiating events." It is indeed hard to invent a reactivity accident, but if we consider the unlikely fault of failing to scram the reactor immediately after the loss of all coolant circulators, we find that the negative temperature coefficient is not enough to shut down the reactor without melting some cans. Since the stainless-steel cans absorb several percent of the reactivity, this event takes on the characteristic of a reactivity accident. A rather extensive core melt out could follow. The effect of this on coolant pressure has been calculated and should not pose a threat to the pressure vessel. The extremely high reliability in the control and shut-off systems ensures that this event is extremely improbable.

Summarizing this review in terms of a safety philosophy, it must first be recognized that none of the various types of power reactor could accept the continued addition of the reactivity which is normally available in control rods. Such a prospect simply leads to ideas of widespread overheating, over pressure, loss of containment, a general picture which we have suggested should be held at a chance of occurrence of less than 10^{-7} per reactor per year.

For the slow reactivity addition rates, the scram system must provide this required reliability, taking due credit for the small chance of the reactivity addition occurring. For the fast rates, the scram system would be too slow; therefore, the case rests upon being sure to this degree of reliability that the initiating event will not occur. If, as in the example of the LWR rod ejection accident, the countermeasures include both design prevention and operational limits, the desired value of 10^{-7} should be attainable.

An inadvertent reactivity addition during start-up is a special case, because several of the trip parameters normally relied upon, e.g., pressure and temperature, may be too far below their design value to be usable. This would reflect on the reliability of the trip function at this time. Against this may be offset the infrequency of start-up, and the fact that an abnormal mode of start-up should be a prerequisite in order to depart so far from equilibrium conditions. In other words, some reliability should be attributable to the sequencing and control rod speed control as designed.

Loss-of-Flow Faults

Coolant is circulated through the core by primary circulation pumps with associated jet pumps in the BWR which consume about 6 MW. Such pumps require their own control and protection,

8. Thermal Reactor Safety

and from time to time fail. A round figure for frequency of failure, a complete rundown of pumps, is once per year per reactor. Flow rundown rate is then mainly determined by the pump inertia; typically, flow halves in about 10 sec. Such rates allow ample time for detection of the condition from multiple signals, the power to the blowers, pressure drop, and coolant temperature, to scram automatically. Transient temperature rises should be trivial; indeed, following scram, there is something of a problem of over cooling. Flow could still be more than half full flow, when reactor power has dropped to afterheat level of 5%. Temperatures are carefully calculated for this transient, mainly for the purpose of evaluating the over cooling thermal shock on the reactor structure.

The prime safety requirement is to achieve sufficient reliability from the reactor scram for this purpose (i.e., 10^{-7} failures per demand); and secondly, to achieve sufficient reliability for the afterheat removal system to cope with the heat after shutdown.

In most cases, a supplementary low-power coolant system whose capacity is approximately matched to the afterheat is provided. If we were utterly dependent on this system, it would mean that the chance of failure should be less than 10^{-7} during any time that it is in use. Obviously, that would be difficult, so we now consider more closely the extent to which we are dependent on these systems.

In the shut-down BWR, no forced coolant circulation is necessary; the channels continue boiling, and the steam finds its way to a dump condenser. In the PWR case, even without forced circulation, the afterheat would be easily transferred by local natural convection to the water in the reactor vessel. If required, no doubt natural convection could be arranged between the core and the secondary heat exchangers and with a normal PWR layout this is the case. Alternatively, we may simply consider that the very slow rate of temperature rise in the reactor vessel gives time to enable the operators to rectify the fault.

The following considerations apply to the loss-of-coolant circulation in a gas-cooled reactor. The reactor must be shut down; the coolant has little heat capacity to take up the stored heat or decay heat; some heat sharing can take place between the fuel and the graphite moderator. After some time heat must be taken from the reactor into the boilers; the transfer could be arranged by natural convection of the coolant at working pressure but not at low pressure. Power is required to provide forced circulation for this latter case, and also to supply feed water into the boilers. The time available to start auxiliary power supplies is a function of the fuel rating and the detailed design and layout of the reactor and boilers, but within this time the provision of power is essential to prevent fuel melting.

Summarizing the safety review of the loss-of-flow fault, there is a general requirement among all reactor types for shutdown (probably scram) and for auxiliary cooling; each of these should be achieved with high reliability. The former is the special aim of the reactor protective system; it would be quite unrealistic to suppose the latter to be achievable if evaluated as an installed automatic system. It can only be achievable if the surveillance and capability of the station staff are given credit, and for this time is required. For this accident, the BWR intrinsically provides the longest time and should require the simplest services. The sodium-cooled fast reactor is in a similar category. The gas-cooled reactors may present special problems of boiler control.

Loss-of-Pressure Faults

Considering a PWR, the pressurizer is provided to compensate for changes of liquid density due to temperature changes, but it cannot compensate for a continued loss of coolant due to a leak in the circuit; this is the function of the make-up pumps. Typically, these will cope with a leak rate of 1000 gal/min; i.e., the discharge through a break of about 10 in.^{2} The process of shutting down and subsequent cooling of the core is simple and well understood for breaks larger than 10 in.^{2} but becomes more complex when the break exceeds 50 in.^{2} or so. Emergency cooling water (ECW) is required, but it is very difficult to specify the quantity, pressure, place, and time of injection to prevent fuel overheating. The LOFT test at Idaho was devised to clarify the ECW requirements; the tests have been difficult and prolonged and are still continuing. Much effort has been devoted to experiments of various types and to the creation of more and more refined mathematical models. A useful review of the whole field is given in [1].

After some time and if the emergency measures are successful, the fuel temperature rise is halted and temperatures fall again as water floods into the core. The total time to quench the core is, therefore, the time for pressure to fall, plus the time to reflood. The former may be typically 20 sec for a full pipe break, or 10 times as long for a break of 50-100 in.^{2} The latter would take about 200 sec. During all this time the heating or cooling of the core is very much in question. Appeal is made to "computer models," but so many effects need to be included that there is currently some disagreement among specialists as to what are appropriate assumptions. The USAEC regulatory staff have published their requirements on what are

8. Thermal Reactor Safety

acceptable temperatures in the transient. This has been fully documented [2]. In brief, the aims are

(1) To keep fuel geometry more or less unchanged, so that the path for recooling is not in question.

(2) To limit the amount of metal/water reaction

$$Zr + 2H_2O = ZrO_2 + 2H_2 + heat$$

partly because of the chance of explosion from the hydrogen, partly because of the heat, and partly because of the cans which become brittle if more than about 10% of the can wall reacts.

Some difference appears to exist in attitudes toward "can ballooning." Zircalloy is rather ductile at temperatures around 1000°C, exhibiting about 40% elongation to failure. Consequently, if the external pressure of the reactor water is removed from a can, which is then held at, say, 1000°C, the internal gas pressure in the can will tend to make the can swell, and it is UK experience that in the temperature range 950-1100°C this swelling can be remarkably uniform. Therefore, within one or two minutes, depending on the temperature, pressure, and can thickness-to-diameter ratio, the cans may press against each other forming a sort of honeycomb which coolant cannot reenter. At higher temperatures, the swelling is more rapid and less uniform; the local swellings offer little extra resistance to coolant flow. In the operation of the prototype SGHWR in the UK, the uniform swelling must be restricted—tentatively to 5% diameter increase. This tends to be the most limiting criterion.

The description of a similar accident on a BWR would deviate in one significant respect. A modern BWR achieves its water recirculation via internal pumps inside the pressure vessel. Only about 10% of the recirculated water is ducted outside the pressure vessel. These loops offer a comparatively high flow resistance; consequently, this type of reactor does not have the low-pressure drop bypasses of the PWR heat exchangers. In consequence, the direction and magnitude of flow through the core is more predictable in the early stages of the accident, and there seems to be more chance of removing the stored heat from the fuel before flow stagnation can take place.

A major step in remedying these uncertainties as to whether the critical conditions are correctly specified and whether the designs meet the criteria has been taken from about the beginning of 1975 by the offer of modified fuel bundles for both PWRs and BWRs, known as the 17 × 17 cluster for PWR, and the 60-pin bundle for BWR. The quantity of fuel in the reactor has not been changed, but it has been subdivided into smaller diameter pins. The advantages will be less stored heat per pin, a greater heat transfer surface area, lower internal fission product gas pressure, and greater strength to resist ballooning. For current power ratings (power per fuel element assembly), it will be much easier to demonstrate good safety margins.

The attempt, in SGHWR, to inject water directly into the core does not entirely obviate the difficulties. At the lower pressures—5 atm or so—the steam formed from the injection spray is inefficient as a coolant. Current designs aim at limiting the rate at which the circuit can depressurize by restricting pipe and header sizes. This gives a longer interval at higher pressures, at which the injected water is quite an efficient coolant. Having removed the stored heat, the efficiency at the subsequent lower pressure is sufficient to prevent temperatures from rising. A similar argument—of limiting the depressurization rate by self-imposed restrictions on pipe and header size—appears to apply to CANDU.

With one exception, this expedient is hardly possible on a pressure vessel water reactor. To subdivide the circuits of a PWR would be very expensive in pumping power and construction cost. The exception is a steam line break on a BWR. These pipes pass from the reactor through the containment to the turbine. If such a pipe broke, or if the turbine exploded, a pair of valves at the containment boundary should close automatically. We should, however, decide what reliance we need to place on the successful operation of the valves. If they did not close, the steam flow down the broken pipe would not be inordinately large—perhaps twice normal. If the steam separators at the reactor still functioned successfully at this flowrate, about 2/3 of the reactor water would flash off into steam, leaving 1/3 at atmospheric boiling point; the core cooling should be good, no major hazard being likely. If, however, much water were entrained in the flashing process, the core could be emptied. If cans fail at this time, the fission products would be carried outside the containment via the broken steam line. Therefore, it is important to know the characteristics of the core depressurization in this accident; if necessary, the steam line probably could be subdivided without important effects on capital cost and efficiency.

However, it is interesting to summarize the factors which would enter into this decision, viz., whether to subdivide the steam mains to the turbine.

(a) The chance of the initiating failure. The chance of a disintegrating type of turbine failure has been assessed in the range 10^{-2}-10^{-3}/turbine yr.

(b) The chance that the containment valves will not close. This may be 10^{-1} per valve per demand, giving 10^{-2} for the pair.

(c) The chance that the core spray (which is provided

8. Thermal Reactor Safety

both on BWR and SGHW) may fail to cool every channel. If this were 1 chance in 100, the total chance is down to 10^{-6}-10^{-7}/reactor yr.

This is the type of probability assessment which has been published in WASH-1400 for typical PWRs and BWRs and which has long been applied by the UKAEA to SGHW and PFR. It is useful to separate two types of problems in the above logic. One is the listing of all the event sequences which would cause a significant fission product release if they all occurred together, e.g., represented by (a) + (b) + (c) in the above example. This is described as "event tree" methodology. Another is the calculation of the expected failure rate for a given system—exemplified by the figures, e.g., 10^{-2} per demand, given above. Usually this is done by analyzing the system into subsystems, components, etc., and assigning a failure rate to each derived from experience. This type of analysis is called a "fault tree."

The gas-cooled reactors do not have the marked transition in heat removal efficiency due to the depressurization phase change from water to steam. The heat removal capacity of the coolant drops in step with the pressure. The working pressure for the AGR is around 40 atm, so that at atmospheric pressure the coolant flow would drop to 2 1/2% of normal. For the pressure to fall rapidly to atmospheric pressure would require a catastrophically large break in the pressure vessel. The concrete pressure vessels now used have composite closures of main penetrations; if one part only failed, the leak rate would be limited, the time to depressurize being about 20 min. At this time the decay heat is less than 2% of normal power.

As explained under "loss-of-flow accident," feed water has to be maintained, as well as blower circulation. However, the expected frequency of depressurization, perhaps 10^{-3}/yr, is less than that of loss-of-flow of once every few years. If feed water reliability is adequate for the latter case, it should also be satisfactory for the loss-of-pressure event. If we are considering a contained reactor system, the loss-of-pressure case may involve the passage of radioactivity from the primary circuit to the containment space which may subsequently prevent access. The reliability of feed and flow may, therefore, have to be considered against a long time scale—say, 6 months or a year.

Summarizing, the expected frequency of a loss-of-pressure event is judged to be lower than 10^{-3}/reactor yr. If the emergency cooling is successful, little activity is released to the containment; one could imagine partial success, with some release to the containment—the overall probability would be quite acceptable, i.e., 10^{-7}/reactor yr—if each of these steps had, say, 10^{-2} chance of failure. If the emergency cooling fell short to a more marked extent, the hydrogen produced by the Zircalloy/steam interaction could pose a threat to the containment—if it entered the containment building and detonated, the building must be expected to fail. Similarly, a large quantity of molten core debris is expected to remain molten due to afterheat, and would ultimately penetrate the floor of the containment. Thus a higher assurance is required against total failures of ECW than against partial failure.

References

1. American Physical Society. Report on LWR Safety, *Rev. Mod. Phys. Suppl. 1* (1975).
2. USAEC. Concluding Statement of the Regulatory Staff, April 16, 1973. USAEC Docket RM-50-1.

9

Safety of Fast Reactors
H. J. Teague

Fast reactors are likely to make a major contribution to the production of electricity in many industrialized countries by the turn of this century. It is important, therefore, that not only should their safety standards be high enough to permit them to be installed near centers of population, but that the general public should accept this assessment. The aim of this chapter is to provide general information rather than to provide detailed estimates of the probability of specific hazards and of the efficiency of the safety measures designed to prevent them or to mitigate their consequences.

All reactors are designed to operate safely and reliably, and incorporate a variety of safety devices intended to ensure that any fault which occurs, even as a result of deliberate maloperation, will not produce a hazard. It is the duty of the safety assessors to consider all the faults which may occur and, adopting an appropriate degree of pessimism towards the safety devices, to consider the results which could follow. In some cases it will be possible to establish that, even assuming that the protective devices fail to function, no hazard external to the reactor will arise (though the reactor itself may suffer damage). In other cases there may be a risk which could be reduced or eliminated, and designers and operators are encouraged to find effective ways of doing this, having in mind the size and severity of the risk in relation to the penalty.

During the period when a new type of reactor is being designed, it is usual practice for the designers and safety staffs to carry out an assessment of its hazards which not

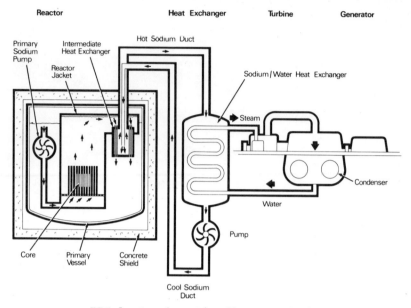

FIG.1 A schematic diagram of the PFR.

only identifies them but sometimes assigns numerical probabilities to their occurrence and estimates the most serious consequence, primarily in terms of the release of radioactivity which could occur outside the site boundary. As the design develops, influenced by safety considerations, the assessments are refined, supported by both theoretical calculations and practical experiments to establish more accurately the numerical probabilities. Nor does the refining process stop with the completion of the design. Safety work is a continuous process, both in relation to increasing the fund of general knowledge and in relation to specific types and individual designs of reactor. It is, happily, an area in which commercial competition does not impede international cooperation.

Although the principles described in this chapter are generally applicable to all sodium-cooled fast reactors, the design solutions referred to mostly relate to the Prototype Fast Reactor (PFR) developed by the UK Atomic Energy Authority and illustrated in Figs.1-3. This reactor is typical of the so-called pool-type liquid metal fast breeder reactors (LMFBRs) in which the entire primary circuit is immersed in a large tank of sodium.

The primary vessel is of 12.7-mm thick stainless steel, 12.2 m in diameter and 15.2-m deep; it is enclosed in a close-fitting thermally insulated leak jacket, and both tank and jacket are suspended in a concrete vault from a support structure spanning the vault. This structure, usually referred to

9. Safety of Fast Reactors

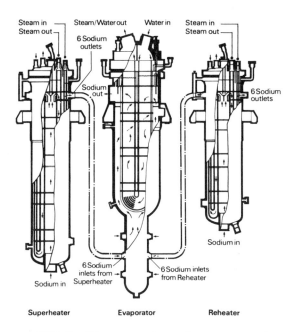

FIG.2 Diagram of a steam generator.

as the "reactor roof," also carries the primary circuit components, thereby eliminating any penetration of the tank below the sodium coolant level and ensuring the highest possible integrity against loss of coolant. In the center of the roof is a rotating shield, carrying the refueling machine. Within the tank is the diagrid support structure. The diagrid itself carries the fuel, breeder, and reflector subassemblies, which are surrounded by neutron shield rods mounted on the support structure.

The pumps draw coolant from the pool of sodium in the main tank and deliver it downwards through a valve to pipes which feed it to the diagrid. From there it flows upwards through the core and breeder assemblies, picking up heat. The hot sodium rises inside the reactor jacket and flows downwards through the heat exchangers, where heat is transferred to the secondary sodium. From the heat exchangers the sodium, now much cooler, flows into the part of the tank outside the reactor jacket and back into the pool. The multiple heat exchangers and pumps give reliability in operation and safeguards against individual failures.

To increase the breeding of new fissile material, the core of enriched fuel is completely surrounded axially and radially by breeder elements and the radial breeder in turn is surrounded

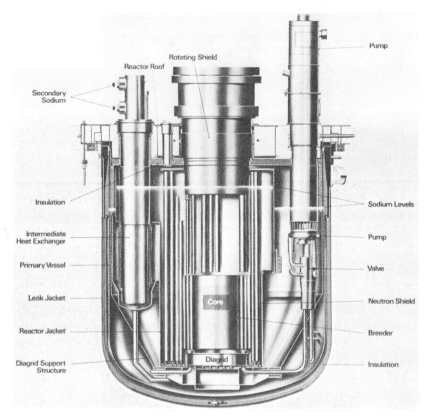

FIG.3 Diagram through the reactor vault showing arrangement of components.

by a steel reflector. The core, breeder, and reflector are all made up of subassemblies, 3.81-m long, of identical hexagonal cross section, 142 mm across the flats. The diagrid supports a number of carriers, each of which has a central "leaning post." Apart from the six around the periphery, each leaning post is surrounded by 6 subassemblies which are tilted slightly inwards and spring loaded against the post. This ensures rigidity of the core. Some of the central tubes act as guides for the control and shut-off rods while others contain experimental assemblies or instruments.

The standard fuel subassembly contains 325 fuel pins, 5.84-mm diameter and about 2.25-m long, enclosed in a stainless-steel wrapper, and supported at intervals by grids. The stainless-steel pins contain a 914-mm length of fuel in the form of pellets of mixed plutonium and uranium oxide. The core consists of two zones containing, respectively, fuel of two different proportions of plutonium to assist in achieving a more uniform power distribution. The inner zone can accommodate 30 subassemblies in which the proportion of plutonium in the mixed oxide fuel is about 20%; the outer zone has spaces for 48 subassemblies containing 25% plutonium.

The secondary circuit ensures the complete physical separation of the steam generators from the reactor and from the radioactive primary coolant. It consists of three separate loops, each loop being connected to two intermediate heat exchangers. After passing through an isolating valve, the hot sodium flow is shared between superheater and reheater sections of the steam generators and is then recombined to pass through the evaporator section. From the evaporator it is returned to the intermediate heat exchanger by a secondary mechanical pump through expansion tanks and a further isolating valve.

The complex nature of the reactor plant and the need to follow demand changes in electrical load require comprehensive instrumentation to maintain control. The main block of instrumentation consists of a data processing system supervised by an on-line computer. This collects information from 1700 analog and 1000 digital inputs and, by means of about 35 programs, processes it to produce:

(a) operator guidance data, which is displayed on cathode ray tubes in the central control room and at various positions in the plant where portable cathode ray tubes can be used;

(b) control information for the two automatic control loops concerned with ensuring correct steam conditions at the turbine in the face of load changes;

(c) a number of sequence control operations, e.g., fueling operations and interlocks; and

(d) records and analyses of plant performance.

To ensure high reliability in performance, the data recording equipment consists of two identical subsystems, each based on a digital computer. These two subsystems will normally operate together, each dealing with about half the measurements and communicating with the other, but either subsystem alone can handle all the measurements and maintain plant operation. Data is displayed to the operators in the control room on cathode ray tubes; the permanent display is restricted to the essential parameters. The operator can additionally call up displays giving all the information of a particular type which may relate to a particular component, such as a pump, or be required for a particular operation. The displays may be numerical data, mimic diagrams, or performance graphs.

Routine control of the reactor, including compensation for changes in reactivity of the fuel, is provided by five control rods made up of tantalum plates and located in five of the leaning posts around the edge of the inner core. Their movement is controlled by the data recording equipment to maintain constant steam temperature at the turbine. Constant steam pressure is also automatically maintained by the adjustment of sodium pump speeds to produce the required variation in heat removal.

In the automatic protective system two independent sets of trip circuits are provided which release the control rods and a separate set of six shut-off rods into the core. The shut-off rods are made up of clusters of stainless-steel cans filled with boron carbide and are located at the edge of the outer core. One set of trip circuits uses relays and the other a solid-state logic design. Both work on the standard two-out-of-three principle.

Because the control and shut-off rods are suspended in guide tubes which pass through the rotating shield in the reactor roof, they have to be fully lowered into the core and detached from their mechanisms during refueling operations. This ensures that no dangerous build-up of reactivity can occur during refueling.

Three design decisions have greatly simplified the provision of shielding and containment for PFR. The decision to have the entire primary circuit installed in a concrete-lined vault below ground level, combined with the provision of internal shielding of the secondary sodium in the reactor tank, meant that radiation shielding was only required in the reactor roof. The choice of sodium as the coolant, having a very high boiling point (880°C at atmospheric pressure), does away with many of the problems which are associated with the use of high-pressure coolants, which need to be contained in thick pressure vessels.

To provide adequate shielding to the working space above the reactor, the roof structure is filled with a sufficient thickness of concrete and the rotating part of the roof, which has a number of penetrations, has added thickness, filled with iron and epoxy resin to increase the effectiveness of the shielding.

Secondary containment is provided by building the main reactor hall to a high standard of leak tightness and providing a filtered ventilation system to control atmospheric release from the building.

Special Features Important to Safety in Fast Reactors

Delayed Neutrons

Most of the neutrons ejected in the fission process appear virtually instantaneously and are known as *prompt neutrons*.

9. Safety of Fast Reactors

The average time which elapses between the birth of a prompt neutron in one fission and subsequent fission by the same neutron is known as the *mean prompt neutron lifetime*. It is very short in fast reactors, about 10^{-6} sec, which may be compared with 10^{-5}-10^{-4} sec in water-moderated, and up to 10^{-3} sec in graphite-moderated reactors. A small fraction of neutrons, however, is emitted at random with mean delays ranging from a few tenths of a second to about a minute, and these are known as *delayed neutrons*.

The delayed neutrons represent a definite fraction of all neutrons produced in fission but this fraction depends on the species of nucleus in which fission occurs. The *delayed neutron fraction* varies from 0.2% for ^{239}Pu to 2.2% in ^{232}Th, and for any reactor there will be a particular value determined by the distribution of fission in the constituent materials in the reactor. A typical value for a plutonium-fueled fast reactor like PFR is 0.3%, though in a ^{235}U/^{238}U system such as the Dounreay Fast Reactor (DFR) the value may be much higher, e.g., 0.75%.

The delayed neutron fraction is of the highest importance to transient reactor behavior. If the reactivity is suddenly increased by an amount less than the delayed neutron fraction, the prompt neutron population grows rapidly, producing a very sharp increase in power. However, the production of delayed neutrons is held up by their inherent delay processes, and the immediate power increase is limited to a level at which the reactivity addition is balanced by the transient relative shortfall in delayed neutron production; from then on power growth is determined by the delayed neutron characteristics.

In consequence, for reactivity changes within the delayed neutron fraction, the kinetic behavior of both fast and thermal reactors is essentially the same, being primarily a function of the delayed neutron fractions. It is often convenient to express reactivity changes as a proportion of the delayed fraction, for which the unit is the dollar. For small reactivity additions each cent produces an immediate 1% power change by prompt neutron increase. Small reactivity additions, provided they fall short of a dollar, produce essentially the same effects in fast as in thermal reactors.

Specific Power

A breeder reactor requires a large mass of fissile material which, for reasons of economy, needs to be employed as intensively as possible so that high power densities in the fuel are an essential feature. Peak specific powers of up to 300 W/g may be employed, and correspondingly high cooling capability is provided, so that in normal conditions, whether shut down or operating, there is no direct problem due to the high rating. In accident conditions involving loss of cooling, the time scale of further accident development is correspondingly shorter than in water reactor systems where typical peak rating may be up to 40 W/g. Totally uncooled fast reactor fuel at maximum normal rating would heat up at rates approaching 1000°C/sec. However, the overall effect of this difference in time scale is not very significant when allowance is made for the wider margin to coolant boiling or burnout in fast reactors. A further consequence is that internal fuel temperatures need to be kept at acceptably low levels by limiting the diameter of fuel pins. The coolant channels which result are correspondingly narrow and, therefore, are susceptible to blockage.

Reactivity Feedback Processes

There are three reactivity feedback effects of paramount significance in fast reactor safety:

Doppler coefficient;
temperature and power coefficient due to thermal expansion of fuel and core structure; and
sodium voidage and temperature coefficient.

All affect the response of the reactor to reactivity and cooling disturbances, and hence influence the control and protection systems, usually reducing demands on them by providing some measure of negative feedback. However, these effects are not important in normal operation; in principle, satisfactory control and protection could be obtained even if the reactivity temperature and power coefficients were nonnegative. The overriding importance of the Doppler and void coefficients in fact lies in their significance in the context of severe accident situations.

The Doppler coefficient acts by the effect on self shielding when the absorption resonances observed in nuclei of even mass number are broadened owing to increased thermal motion. The resonances occur in a region of the neutron energy spectrum which contains only a small proportion of neutrons even in large fast reactors, for which the lower enrichment leads to a relatively softer spectrum. In small reactors the spectrum may be so hard that no appreciable Doppler coefficient exists. Calculation of the effect is complex and difficult for a number of reasons: the effect is fundamentally second order; the tail of the neutron spectrum is concerned; a large number of resonances has to be considered, including the effect of fission resonances in nuclei of odd mass number; and interaction between neighboring resonances is appreciable. Experimental verification is difficult owing to the physical problems of heating the assembly and eliminating thermal expansion effects. Many years of worldwide effort have now resulted in an ability to predict coefficients with a standard error of 25%. The basic physical mechanism of the Doppler coefficient is the thermal agitation of the nuclei and so it is virtually instantaneous in operation. Furthermore, it does not depend on the mechanical integrity of the fuel. Thus, it constitutes a feedback mechanism which can be relied upon even in the context of a super prompt critical excursion. It is, in fact, found that the Doppler effect is a major factor in limiting the energy released in such an excursion.

In any reactor small movements due to thermal expansion produce reactivity changes. Such movements may be rather complicated, including bowing effects as well as uniform expansion; in general, there is also an associated time constant depending on thermal and mechanical inertia.

The sodium void effect also results from a combination of physical processes.

(i) Removal of sodium enhances neutron leakage, thereby reducing reactivity.

(ii) Removal of sodium reduces neutron absorption in sodium, thereby increasing reactivity.

(iii) Removal of sodium hardens the spectrum which increases the "neutron effectiveness," thereby increasing reactivity.

(iv) Removal of sodium reduces background scattering, thereby increasing self shielding in heavy elements, which leads to reduction in effective cross sections, so modifying reactivity.

The overall voidage effect is the result of these four components. Since their relative magnitudes depend on their position in the reactor (e.g., at the edge leakage effects are relatively more important than spectrum effects), the voidage coefficient varies markedly from point to point. It is, in general, positive near the center, decreasing to strong negative values at the edge. For a small core, total loss of sodium will produce a net reduction in reactivity, but for very large cores of conventional design, there could be a net increase. In fact, most practical designs give a negative effect for complete voidage.

Removal of sodium, particularly from the central region of a reactor, is therefore a possible mechanism for rapid reactivity addition. It is important to examine this possibility in relation to passage of a large volume of gas through part of the core, and in the event of an accident resulting in general boiling of coolant.

Sources of Hazard

The steady operation of a reactor at a set power level depends on maintaining a balance of reactivity and a balance of heat removal from the fuel. Any event which increases the reactivity or reduces the heat removal is therefore a potential source of hazard. For some of these events, such as the accidental withdrawal of a control rod causing an increase in reactivity or the failure of a pump leading to a reduction in the rate of heat removal, the response of a fast reactor is very similar to that of a thermal reactor and the well-established safety provisions of the latter can be applied to fast reactors.

For some other events, the response of a fast reactor is sufficiently different to warrant particular consideration. The principal differences which affect the response (not necessarily adversely) are: the much higher volumetric heat rating arising from the higher fuel rating and the more compact core; a prompt neutron lifetime of about one-thousandth of that in a thermal reactor; a coolant effectively unpressurized at a temperature well below its boiling point; a chemically reactive coolant; a core which is not in its most reactive configuration; and a fuel with a high plutonium content.

Apart from considering faults which may occur while the reactor is at power, it is also necessary to consider faults which could affect the reactor when it is shut down and faults which could arise in the handling of fuel into and out of the reactor.

Heat must be transferred from sodium to water, and in current designs of fast reactors there is an intermediate sodium circuit which physically separates the steam generators from the reactor. It also has the advantage that the sodium which passes through the steam generator is not radioactive. The reaction between sodium and water can be violent. To reduce the likelihood of this, great care is taken in designing and inspecting heat exchangers and additional steps are taken to protect the reactor from damage should a leak occur in the steam generator. In most designs there is provision for early identification of any leak with means for isolating the steam generator from the rest of the circuit and dumping the sodium into tanks provided for this purpose. Finally, as sodium burns in air, special provisions are made to reduce the chance of sodium leakage and to limit any fire which might arise. This is referred to later.

9. Safety of Fast Reactors

Faults Producing Increases in Reactivity

It has been pointed out that for small or slow increases in reactivity the response of a fast reactor does not differ materially from that of a thermal reactor and similar methods of detection and shutdown can be adopted. The principal means by which a large, rapid increase in reactivity might be produced are:

(1) sudden removal of absorber;
(2) rapid insertion of fuel;
(3) introduction of moderating material;
(4) change in coolant density due to void effects; and
(5) rearrangement of the core into a more reactive state.

These are discussed in the following paragraphs, together with the measures which can be taken to eliminate them. In all cases, the negative Doppler effect will work toward reducing increases in reactivity as the fuel temperature rises, and would be capable of annulling the effects of rapid additions of reactivity even in excess of one dollar without producing extensive fuel damage.

Sudden Removal of Absorber The only feasible cause of this is the violent ejection of a control rod from the core, and the safest way of countering it would be to ensure that the reactivity controlled by one rod (the control-rod worth) is less than a dollar, so that the loss of a rod would not make the reactor prompt critical. This, however, would require more rods than can be accommodated in the small, compact core of PFR, though it might be possible in a commercial fast reactor with an electrical output upwards of 1000 MW. In PFR, mechanical design features have been incorporated which, assisted by gravity, ensure that the chance of ejection is very small.

The chance that accidental or even deliberate withdrawal of a control rod could cause such an effect is prevented by mechanical control of the maximum speed of movement, which provides ample time for the automatic shutdown mechanisms to operate before a dangerous situation can arise.

Rapid Insertion of Fuel This could arise from a faulty handling machine dropping a subassembly into the core or, less rapidly, from the accidental loading of a subassembly into a previously vacant position, or from the loading of a subassembly of the wrong enrichment. In PFR, it is physically possible to insert fuel into the core or remove a control rod only when the reactor is shut down with all the absorber rods fully inserted into the core. This will provide a sufficient margin of negative reactivity to prevent any of these possibilities leading to criticality.

Introduction of Moderating Materials It is recognized that if moderating material such as oil were allowed to enter a fast reactor, it would lead to an increase in reactivity, and on this account strict control will be maintained over the amounts of such material in the reactor area. In one investigation, to get an idea of what would constitute a dangerous quantity in the core, tests were carried out in the zero-energy reactor ZEBRA using polythene to simulate the oil. When the results were adjusted to compensate for the difference in moderating effect, it was estimated that for there to be any possibility of a prompt critical excursion, at least several hundred grams of oil would need to reach the core in less than one second. Analysis of fast reactor designs has shown that this can be prevented with a high degree of reliability, independent of the protection provided by "good housekeeping," and the hazard from such a fault is not regarded as significant. Efficiency of heat transfer processes would be impaired at a much lower level of contamination than that described above, and consequently demands correspondingly stricter control.

Change in Coolant Density Generally speaking, the net effect is negative near the edges of the core and positive in the central region. The total effect over the whole core tends towards an increasing positive effect as core sizes increase.

The change in density due to a rise in temperature will be too small to have a significant effect on the safety of the system as long as the sodium remains liquid; if the sodium boils, it is presumably due to a preexisting fault condition, such as loss of pumping power followed by failure to trip the reactor, and the question then is whether a positive void coefficient will materially increase the severity of the accident. No general answer can be given and attempts must be made to assess the likelihood in individual cases, but even then the problem is extremely complex and in the present state of knowledge involves an element of judgment. UK studies based on reactors with electrical outputs of 1000 MW indicate that the positive void coefficient resulting from a design optimized for power production does not unacceptably reduce the safety of the reactor. Furthermore, the advantage of a less positive void effect in a core designed for that purpose has in practice to be weighed against an associated reduction in Doppler coefficient.

Other sources of voids in the coolant are bubbles of entrained cover gas and gaseous fission products released as a result of fuel can failure. It would require a large bubble of gas to be swept into the reactor core to cause any significant danger and the risk of its occurrence can be made very slight by suitable design. For PFR, extensive model tests

9. Safety of Fast Reactors

were undertaken to ensure this. Moreover, the presence of very small amounts of gas is readily detected and gas concentrations can be accurately measured. The present limit of detectability is 1 in 10^{10} in sodium loops and similar applications in reactors should readily achieve 1 in 10^7 parts by volume.

As regards the release of fission gas, calculations have established that even the complete voiding of a single subassembly will not produce a hazard as a result of the reactivity effects due to the coolant void.

Rearrangement of the Core Fast reactors differ from many thermal reactors in that a fault which causes the fuel to melt may lead to an increase in reactivity if the molten mass assumes the most compact geometry. The reactivity changes which could occur as a result of extensive and arbitrary rearrangement of sodium, fuel, and steel in a complete subassembly have been calculated in detail, and the calculations have been checked by experiments in the zero-energy critical assembly ZEBRA. The results indicate that the normal operation of the reactor temperature coefficients could cancel any such reactivity addition before damaging temperatures were reached in the rest of the core. The process of damage propagation to a neighboring subassembly by the overheating effects of molten fuel is delayed by the time required to melt through the two wrapper walls which are separated by several millimeters of stagnant sodium. The second wrapper wall, moreover, is well cooled on its inner face by the normal coolant flow. Meltthrough, therefore, is not a mechanism which can directly cause simultaneous rearrangement of large amounts of fuel.

Faults Due to Local Loss of Cooling in a Subassembly

The high heat rating of fast reactor fuel means that sudden loss of cooling, even over a small region, will lead to a rapid rise in temperature. Such a sudden loss could only arise from a blockage of the coolant passages between the fuel pins, and the fuel subassembly has been designed to minimize the risk of a blockage occurring. Filters at the coolant inlet ensure that it cannot be caused by material entrained in the coolant stream, and spacer grids at intervals will prevent distortion of the fuel pins which might otherwise lead to areas of coolant starvation. This leaves the remote possibility of a major failure of a can allowing fuel material to escape and cause an obstruction, although experience, e.g., in DFR, indicates that when stainless-steel-clad oxide pins fail, the mode of

failure is not one in which fuel could be released in a form
or quantity which could block a coolant subchannel. A local
blockage causes only a very small disturbance to the bulk subassembly flow, so that the bulk mean outlet temperature, as
measured by the outlet thermocouple, would not indicate the
presence of a blockage. Local overheating near the blockage
tends to persist in the coolant stream, leading to an increase
in temperature noise as minor fluctuations in flow paths occur,
and it may be possible to achieve early detection by analyzing
the noise spectrum of a very fast response thermocouple. Warning would also be given by the presence of fission products in
the coolant, both through the burst can detection equipment
provided for every subassembly and, more rapidly, but with less
sensitivity, through detectors sampling the bulk reactor outlet
flow. In the highly unlikely event that the reactor was not
shut down, the increasing temperature would lead to sodium boiling and in PFR acoustic detectors are installed to pick this
up. In the later stages, when local boiling escalated to general bulk ejection of sodium from the subassembly, the outlet
thermocouple would register rapid but marked temperature fluctuations in the sodium leaving the subassembly. At the same
stage coolant ejection would be associated with reactivity
changes due to sodium voidage effects, although their variation
with position in the reactor implies that not all subassemblies
would necessarily be covered in this way. Direct flow measurement would give clear indications of substantial disturbance
in the bulk flow.

Thus, a variety of warnings of a blockage are available
in principle, offering a high probability that the reactor will
be shut down before severe damage to the subassembly occurs.
The main practical problem is to ensure adequate coverage of
the whole core without incurring an intolerable burden of spurious trips. Individual subassembly instrumentation, such as
thermocouples or flow meters, is particularly prone to this
drawback, which is why emphasis is placed on "field" effects
such as acoustic and reactivity noise and detection of escaping
fission products.

There remains a slight chance that, in spite of such a
system of local blockage detection, this type of incident could
progress to unstable bulk boiling in the subassembly. In that
event the coolant would flow almost entirely as low density
vapor with a heat removal capability less than half the normal
value with liquid coolant. General fuel melting would follow
within several seconds.

In order to penetrate the wall of the neighboring subassembly, molten fuel would first have to melt through the wall
of the incident subassembly. This would require many seconds,

9. Safety of Fast Reactors

taking account of the heat capacity of the steel, the heat removal capability of stagnant sodium between the wrappers, and the quantity of fuel (limited by melting and slumping) which could remain in contact with the wall. Then it would be necessary to melt the neighboring wall, which would be well cooled by the undisturbed flow of the unaffected subassembly. Throughout this extended period there would be extremely large and unambiguous fission product signals from exposed fuel, leading to safe shutdown.

A further possibility, within the incident subassembly, is that of explosive vapor formation if, as a result of geometry changes at melting, liquid sodium should come into contact with molten fuel. An absolute upper limit to the mechanical energy released can be set by applying fundamental thermodynamic considerations to the process. In principle, approximately one third of the heat stored in the molten fuel could be converted to mechanical work. For a PFR subassembly, the absolute upper limit is, therefore, about 20 MJ.

There are several reasons why the energy released should be much less than the theoretical maximum. A high-pressure incident demands very high rates of heat transfer which can be obtained only if the fuel is finely dispersed and intimately mixed with liquid sodium. Even then, the growth of a vapor film between the two liquid phases might be expected to impair heat transfer and further reduce the generation of mechanical energy. It is unlikely that the necessary conditions for a vigorous molten fuel/coolant interaction would arise in more than a small fraction of the fuel; indeed, some workers believe that they cannot arise at all in reactor conditions. Moreover, irradiated fuel has a high gas content, and it is well established that moderate quantities of gas can inhibit vigorous interactions even between materials such as molten aluminum and water.

Nevertheless, a vapor explosion is a complex process which is influenced by many factors. Some are inherent in the materials such as their physical properties; others are determined by the local environment such as dynamic constraints and mode of bringing into contact. The possibility of energy release of a few megajoules cannot yet be entirely dismissed, although the extensive research currently being pursued has excellent prospects of refining practical limits.

It remains to consider the consequences to the reactor of a fuel/coolant interaction. The strength of a wrapper is limited, particularly so if it becomes overheated in the loss-of-cooling phase of accident development. The surrounding subassemblies do, however, provide considerable resistance to a central pressure and, if they are deformed, the texture of many narrow pins separated by sodium is well-adapted to energy absorption. Consequent further hazards to be considered would be: direct reactivity additions due to core distortion; inability to shut down the reactor because core distortion has inhibited movement of absorber; and crushing of neighboring subassemblies to a degree that prevents heat removal (even at decay heat levels).

Reactor physics calculations generally confirm the commonsense expectation that a dispersive event like a subassembly fuel/coolant interaction is likely to reduce reactivity. Nevertheless, the possibility of reactivity additions should be examined with care in any new reactor, particularly in respect of off-center incidents. Although the obvious risk is that, associated with maximum energy conversion, there may be other hazards peculiar to the less severe reactions. For example, in a freestanding core with wide clearances, a relatively mild pressure pulse may initiate oscillatory movements if the elastic limit of the subassembly support system is not exceeded and rapid reactivity increases could, in principle, occur on the inward part of the cycle.

Experiments with clusters of subassemblies, using rapid gas generation to simulate a vapor explosion, have improved understanding of core distortion processes. Within the expected range of time scale for pressure build-up, neither peak pressure nor energy release are the principal determining factors; of more fundamental significance is the integrated impulse exerted by the pressure pulse. Large fast reactors, in general, can be shut down by inserting a small number of rods (typically 5 or 6) virtually independent of their position in the core. Since there would be a much larger number of rods (typically 30-40) widely distributed through a large fast reactor, quite severe local damage could be tolerated because the reactor would be shut down by control rods operating in regions remote from the damage. PFR can be shut down by the insertion of any 3 of its 11 rods.

Distortion of neighboring subassembly geometry on the whole appears unlikely to be so severe that sodium circulation would be impaired to the extent that decay heat could not be removed without melting fuel. Naturally, the mechanical characteristics of the core strongly influence its response to a fuel/coolant interaction. A freestanding core, such as PFR, is likely to suffer little crushing but fairly extensive displacement, so that interference with control rod movement is probably more important than loss of cooling capability. In a highly constrained core the reverse might apply.

During many years of experimental fast reactor operation, considerable experience has been accumulated of fuel behavior.

9. Safety of Fast Reactors

Low pin-failure rates have been established, but there have been some instances of extensive pin failure. In none of these, however, did unstable flow conditions arise, nor has there ever been any evidence of explosive vapor generation. It is also noteworthy that despite an extensive and prolonged experimental program, it has been extremely difficult to induce such reactions in the laboratory. Those which have occurred took place in special circumstances on a strictly limited scale.

One hazard arising from molten fuel, produced in a subassembly accident, for instance, is that it might melt through the core support structure, collect at the bottom of the reactor tank, and melt through that. In PFR, this is prevented by the installation of a collecting tray below the core designed to disperse the molten debris so as to avoid the formation of a critical mass, and to permit cooling by the primary circuit sodium. Calculations indicate that it could successfully retain half the complete contents of the reactor core, at heating rates dependent on the decay of fission products. The question of a more extensive melting of fuel, such as might lead to a prompt critical condition, is examined in more detail below, under the heading of the effect of a "hypothetical" severe accident on containment.

Fuel Handling Faults

The principal potential hazards of fuel handling operations arise from

(1) radiation from irradiated fuel;
(2) decay heat in irradiated fuel;
(3) release of fission products following a fuel pin failure;
(4) radioactivity, chemical activity, and flammability of the coolant; and
(5) criticality.

The faults which may activate these hazards will depend on the methods used in reactor refueling. The generally accepted view at present is that for fast reactors the complexity of providing on-load refueling is not justified in view of the relatively small number of handling operations involved. For PFR, the handling between the reactor core and the stores of new and irradiated fuel takes place in two stages. The intermediate location of the fuel is a storage rotor submerged in the pool of sodium in the reactor tank, and fuel can be transferred between this rotor and the fuel stores while the reactor is operating at full power. For transfers between the rotor and the reactor core, a charge machine mounted on the central rotating section of the reactor roof is used, and this can only be operated with the reactor shut down and all absorber rods fully inserted in the core. As all the handling operations between core and rotor take place with the fuel subassembly submerged in sodium, the removal of decay heat is not a problem, even if the operation is delayed at some point owing to loss of power supplies or breakdown of the refueling machine. Damage to a subassembly or to the reactor internals could occur during refueling, however, if the subassembly were dropped because of mechanical failure of the grab, or if operations were carried out in the wrong order. These risks can be made very small by good design and by providing a system of interlocks or of sequence control by computer. Such accidents, even if they occurred, would not produce any hazard outside the reactor.

The decay heat from a fully irradiated subassembly is about 600 kW at shutdown, falling to about 120 kW after 3 hr, which is considered to be the minimum time required to prepare the reactor for refueling. From the point of view of safety and ease of handling, the subassembly should remain in the storage rotor as long as possible; a 30-day cooling period would reduce the decay heat to not more than 15 kW. To transfer the irradiated fuel from the storage rotor to the irradiated fuel store, it is drawn up a discharge tube passing through the reactor roof shield in a bucket of sodium into a shielded transfer flask. With a suitable design of bucket, the decay heat can be removed by forced convection and radiation from the bucket walls, which may be finned to improve heat transfer. As a further safety measure, the walls of the discharge tube may be cooled, to guard against a delay caused by a fault in the hoist. Further precautions against possible faults in this part of the transfer are the provision of a shock absorber at the bottom of the discharge tube, to minimize damage if a grab failure releases the bucket, and a sensor to check that the sodium is not leaking from it. The risk of a leak is extremely slight and can be reduced by periodic inspection; the risk that a leak occurs and is not detected is even slighter, but such a fault could lead to overheating of the fuel, failure of the cans, and release of gaseous fission products. Even if this occurred between the reactor vessel and the fuel store, the flask is intended to prevent their dispersal, which would in any case be into the secondary containment and the risk of external hazard is acceptably low.

The irradiated fuel store must be designed to provide protection against the activity of the fuel, cooling to remove the decay heat, and sufficient dispersal to assure against criticality. All these can be provided with the requisite reliability by careful design.

9. Safety of Fast Reactors

Containment

The purpose of containment is to provide a barrier against the dispersal of radioactive material which may be released under fault conditions. The great bulk of radioactivity is in the fission products and in most reactors, including PFR, the fuel cladding provides an initial barrier to their escape; a large release of activity from this source must, therefore, presuppose that very many cans have failed, which is an improbable event. Other possible, though relatively minor, sources in PFR are the active sodium and the reactor cover gas. Thus a very high percentage of the active material is contained within the reactor tank, and the vault and shielded roof provide a primary containment which is intrinsically very strong and capable of withstanding a range of accidents up to a degree of severity well in excess of that which can be foreseen as arising even from a complete failure of the protective system. These severe but hypothetical accidents are considered below.

Outside the primary containment, possible sources of radioactive releases are accidents during the transfer of irradiated fuel and leaks in the active sodium purification loops. The logical boundary for the secondary containment is the reactor building, and this must be capable of coping with the highest release of activity which could result from these faults, even though access for personnel and transport must be provided. This capability has been achieved by the design of the building structure and the provision of extraction pumps which discharge air from the building through filters to a chimney and can maintain an internal pressure slightly below atmospheric inside the containment.

This makes leakage paths less critical, since the leaks will generally be inward, and does not require the structure to be strengthened to withstand internal pressure. It also forms a useful additional protection against a wide range of accidents which may result in some leakage through the primary containment.

The Effect of a "Hypothetical" Severe Accident on Containment

It will be seen that the basic approach to fast reactor safety is to provide instrumentation and protection which give an extremely high degree of assurance that in spite of malfunctions or failures in any part of the plant, the core will be preserved in a substantially intact state. The basic stability of fast reactors and their wide margins between operating and damage conditions facilitate the attainment of high standards.

At worst, given that the protective systems act as they are designed to act, the containment merely has to prevent the escape of activity from limited amounts of damaged fuel in conditions of temperature and pressure which are quickly restored to normal. However, no system is perfect and there is no guarantee that all types of accidents have been considered, so that it is common to study the consequences of failure of plant and equipment even if the failure in itself is highly improbable.

An extreme instance of such a severe accident arises when the core becomes prompt critical. Analysis shows that what is important is not the total amount of reactivity added, if this is of order 1 1/2-2 dollars, but the rate of addition when prompt criticality is attained. It is unrealistic to specify a particular reactivity ramp rate, because the physical processes acting are varied, complex, and not amenable to precise analysis. It is useful, nonetheless, to study typical processes, such as coolant expulsion and slumping of molten fuel under gravity, in order to gain some appreciation of what rates are feasible.

It is found that material movements under gravitational forces, pump pressures, and bulk sodium boiling may cause reactivity to increase at rates in the range \$10-50/sec. Larger forces are conceivable, such as those from concerted release of molten fuel leading to an unfavorable pattern of fuel/coolant interaction which could theoretically lead to reactivity increases at a rate of a few hundred dollars per second. Current indications are that if fuel pins fail in the mode which is believed to be correct, this unfavorable pattern will not occur and reactivity changes will be small or even negative because fuel will be swept out by expanding vapor. As described earlier in this chapter, the Doppler effect is extremely important in providing a fast negative feedback in reactivity which keeps the reactor close to prompt critical with the temperature rising at a rate sufficient to counter the applied ramp. For small ramps this process of automatic adjustment is smooth and continuous, with the reactor maintaining itself just below prompt criticality. For larger ramps power is generated in surges, with the temperature rising in steps and the reactivity fluctuating around prompt criticality. The process continues until the energy density in the fuel gives rise to pressures which will dilate the core, and cause a permanent reduction in reactivity to a level well below delayed critical. When this stage is reached, the energy density in the fuel has attained a final value which depends in a rather complicated way on the ramp rate, the Doppler coefficient, and the effective equation of state of the fuel. In general, the energy density is high enough to result in a vigorous dispersal of the fuel.

9. Safety of Fast Reactors

On the conservative assumption that dispersive pressures are those due to UO_2 vapor pressure, it is obvious that a considerable proportion of the calculations would imply that the central pressures may be extremely high—several hundred bars for a ramp of a few hundred dollars per second—and that the potential for performing mechanical work may be in the range of hundreds of megajoules depending on the size of the reactor.

The direct effect of an energy release of this type is to send out pressure waves, similar to mild shocks, and to accelerate the surrounding sodium to high velocities. Consequent hazards are damage to the roof of the primary containment caused by the impact of sodium and damage to the intermediate heat exchangers and emergency decay heat removal systems. The primary containment is intrinsically very strong, and the structures surrounding the core, particularly the breeder and neutron shields, provide good protection for heat exchangers and the primary vessel. The actual level of energy release which can be withstood will depend on detailed design characteristics, but can be made high, in excess of 1000 MJ.

A less direct effect may be the generation of sodium vapor due to contact between the surrounding coolant and the large mass of molten fuel. The total quantity of heat could be absorbed by the sodium without vapor production provided that the heat is able to penetrate far enough into the bulk sodium. If limited quantities of sodium are involved, large volumes of high-pressure vapor could be generated. Apart from the immediate dynamic effects, sufficient vapor could be produced to pressurize the primary containment in a quasistatic manner, since recondensation would be slow in the presence of blanket gas. It is, however, practicable to provide sufficient free volume to accommodate sodium vapor at a pressure within the strength of the containment.

Fast reactor fuel at equilibrium contains a large quantity of volatile fission products which may reasonably be assumed to be released on fuel melting. Taking into account the available free volume within the core and the proportion of fission gas diffusing to the fission gas plenum, it is possible to estimate the potential dispersive pressure which would be generated when the fuel melts. If this is used in core supercritical excursion calculations, the character of the predicted energy release undergoes a marked change. The proportion of fuel raised to boiling point is greatly reduced, and may even be eliminated altogether. Although the pressure from fission products appears to be adequate to disperse the core, it has very much less capacity for sustained mechanical work than an equivalent amount of boiling UO_2 in flashing-off conditions, so that the direct explosive effects are virtually eliminated. The energy present as sensible heat is appreciably reduced, although it remains high enough to produce large quantities of sodium vapor. In assessing the risk from that process, however, the inhibiting effect of permanent gas on heat transfer should be taken into account. Provided that fission gases can be shown to act effectively and on the appropriate time scale, they may be assumed, for ramp rates up to a few hundred dollars per second, to constitute an important mechanism for limiting supercritical energy releases.

As was pointed out in the context of the subassembly accident, fuel debris is capable of melting its way through the bottom of the reactor tank, thus breaking the containment boundary. A fuel debris support system ("internal core catcher"), cooled by circulating sodium, can give a high degree of protection in this respect. Such a device has been provided in PFR. Internal core catchers probably cannot be designed to accommodate, with high reliability, a whole core, particularly following a violent dispersal of finely divided fuel in suspension in sodium. As a protection against limited localized quantities (a few subassemblies), there seem to be good prospects of containing debris which would otherwise be likely to melt through the bottom of the reactor tank. Even in a major whole-core accident, only a fraction of fuel would melt because of the power distribution in the core. Alternative, or additional, protection can be obtained by providing efficient cooling on the exterior of the lower primary containment surfaces.

Decay Heat Removal

The activity of the fission products contained in reactor fuel means that after shutdown there will be a continuing but gradually decreasing production of heat from their radioactive decay. As an indication of the size of the problem, Table I gives estimates of decay heat in a fast reactor, as a percentage of full power, and in PFR in megawatts, at various intervals after shutdown. Unless this heat is removed, temperatures within the reactor will rise, producing results similar to a loss-of-coolant accident. In normal circumstances the heat will be removed by continued circulation of the sodium, but provision must be made for situations involving pump failures or loss of power supplies. However, the 900 tons of sodium in the reactor pool of PFR represent a valuable heat sink and even without any heat removal the temperature would only rise by about 100°C in 4 hr, so that a system capable of removing 1% of full-power heat, that is, 6 MW in PFR, is acceptable. At these levels it becomes practicable to make use of natural convection in order to minimize dependence on power supplies. PFR has a natural circulation liquid metal loop from which the

9. Safety of Fast Reactors

TABLE I
Variation of Decay Heating during
First Day after Shutdown

Time from shutdown (hr)	Decay heat	
	As % of reactor power	In PFR (MW)
0	7.6	45.6
1	1.6	9.6
4	1.03	6.2
12	0.73	4.4
24	0.60	3.6

heat is rejected by forced air convection in heat exchangers located in the walls of the reactor building.

Sodium/Water Reactions

Apart from accidents leading to a release of radioactivity, the fault which probably has the gravest potential economic consequences is a failure in a steam generator leading to a leak of water or steam into the sodium. This will give rise to the production of a large volume of hydrogen, a large amount of heat energy, and corrosive compounds. In PFR, the steam generators consist of three units, each comprising an evaporator, superheater, and reheater. Each of these nine components is essentially similar in design, as illustrated in Fig.2. For example, an evaporator is a long, cylindrical vessel with a header near the top with a vertical central partition. Water enters on one side and flows through a series of U-tubes, emerging on the other side, from where it passes to the superheater and reheater through pipes feeding a mixing chamber in the base. From here it rises in an annular space around the tube bundle to an entry port below the header, flowing concurrently with the water and emerging from an exit port below the other half of the header.

Because of the near certainty of occasional tube failures, extensive experimental work has been carried out at Dounreay to investigate the development of a leak, including tests on full-size tubes in the Super-NOAH test rig. Quantities of sodium and water in the test were more limited than those available in a steam generator. The transient high pressures resulting from the reaction produced no damage or visible distortion in the test vessel, and an effluent clean-up system, based on that designed for PFR, proved successful in the safe discharge of the hydrogen evolved.

The production of hydrogen and an increased pressure in the sodium are both symptoms of a leak which can be used to operate a reactor trip and to initiate the closing of valves to stop the supply of sodium and water or steam to the affected generator.

Fire Hazards

Although sodium has many desirable properties as a reactor coolant, its use does introduce an additional fire hazard. Its ignition temperature in air is 200°C for bulk amounts, such as a pool or solid jet, and considerably lower for a spray of droplets. This means that a significant leak in a coolant circuit, in which the sodium temperature will be considerably higher, could lead to a fire, and this will be accompanied by the evolution of dense clouds of sodium oxide fumes. The burning of 8 kg of sodium in a building of one million cubic feet, assuming that 40% is converted to airborne oxide, would reduce visibility to zero. It is thus very important that a sodium fire should be detected as quickly as possible, and while even a small leak will rapidly become apparent as a result of the cloud of oxide produced, it is desirable to have detectors, preferably of the type which responds to smoke or fumes. In addition to providing for detection and location, there are various design precautions which can be taken to minimize any fire which occurs.

These precautions include the provision of dump tanks on all sodium circuits into which the complete sodium content of the circuit may be directed; the subdivision of the floor area under sodium pipes and vessels to reduce the spread of sodium; and the installation of perforated steel plates, a few inches clear of the floor in these areas, so that the sodium will fall through into a restricted air space, to inhibit combustion.

However, while it is necessary to provide well-planned systems for detecting and extinguishing sodium fires, the surest protection must lie in reducing the probability of a leak by a rigid adherence to a high standard of manufacture and assembly of sodium-bearing components.

One feature of such an approach is the double-walled structure of sodium-bearing vessels and pipes; the space between the walls is filled with inert gas and instrumented with devices to detect the presence of sodium.

Summary

Liquid metal cooled fast breeder reactors in normal operation are stable and reliable; safety studies of fast reactor response to standard faults follow a similar course to those undertaken for thermal reactors. In general, safety margins

9. Safety of Fast Reactors

are found to be wide, and a reliable, effective automatic protective system can be engineered.

Nevertheless, no system is perfect, and it is also necessary to examine the consequences of accident progression beyond the point of fuel damage, however improbable it may be. There are many research programs, in many different countries, directed towards improving understanding of these more extreme stages of accident progression. In the course of the next few years, it will be the aim of such research to resolve the outstanding uncertainties so that the overall risk associated with fast reactor operation can be reliably assessed. This must be done in the perspective of comparable risks associated with other technological developments and freely accepted by contemporary advanced societies.

References

1. MacFarlane, D. R. An Analytic Study of the Transient Boiling of Sodium in Reactor Coolant Channels, Argonne National Laboratory, ANL 7222 (1966).
2. Siegmann, E. R. Theoretical Study of Transient Sodium Boiling in Reactor Coolant Channels Utilising a Compressible Flow Model, Argonne National Laboratory, ANL 7842 (1971).
3. Tilbrook, R. W. Coolant Voiding Transients in LMFBR's, CONF 71302. New Developments in Reactor Mathematics and Applications, Idaho Falls, March 1971.
4. Noyes, R. C. A Discussion of Methods for Theoretical Analysis of Coolant Ejection in FBR's, Including a Description of the ETNA Code, CREST-ENEA, RT/ING-(68)10 (1968).
5. Schlechtendahl, E. G. Coolant Boiling in Sodium Cooled Fast Reactors, Gesellschaft für Kernforchung MBH, Karlsruhe, KFK 1020, EURFNR 701 (1969).
6. Kirsch, D., and Schleiskiek, K. Flow and temperature distribution around local coolant blockages in sodium cooled fuel elements, International Seminar on Heat Transfer in Liquid Metals, Trogir, Yugoslavia, 1971.
7. Gosman, A. D., Herbert, R., Patankar, S. V., Potter, R., and Spalding, D. B. Prediction of coolant flows and temperatures in pin bundles containing blockages, International Meeting on Reactor Heat Transfer, Karlsruhe, October 1973.
8. Brook, A. J. Local boiling in the fast reactor subassembly environment, International Meeting on Reactor Heat Transfer, Karlsruhe, October 1973.
9. Buchanan, D. J., and Dullforce, T. A. Fuel-coolant interactions—small-scale experiments and theory (paper pre-

sented at the Second Specialist Meeting on Sodium/Fuel Interactions in Fast Reactors, Ispra, Nov. 21-23, 1973) Paper No. CLM-P362.
10. Duffey, R. B., Clare, A. J., Poole, D., Board, S. J., and Hall, R. S. Measurements of transient heat fluxes and vapour generation rates in water, *Int. J. Heat Mass Transfer 16*, 8 (1973).
11. Fauske, H. K. On the mechanism of UO_2/sodium explosion interaction, *Nucl. Sci. Eng. 51*, 95-101 (1973).
12. Teague, H. J. Summary of the papers presented at the CREST meeting on fuel-sodium interaction at Grenoble in January 1972 and Report of conference papers 38 a-k on fuel-sodium interaction (also papers 38 a-k), Proceedings of International Conference on Engineering of Fast Reactors for Safe and Reliable Operation, Karlsruhe, Oct. 1972, pp. 812-990.
13. Brook. A. J. Some preliminary considerations relating to an equation of state for irradiated nuclear fuel, *Nucl. Saf. 13*, No. 6, 466-477 (1972).
14. Teague, H. J., and Mather, D. J. Factors limiting prompt-critical excursions in irradiated fast reactor cores, *Nucl. Saf. 14*, No. 3, 201-205 (1973).
15. Stevenson, J. M., and Hardiman, J. P. MOZART sodium removal experiments and their interpretation, Proceedings of the International Symposium on Physics in Fast Reactors, Tokyo, Oct. 1973.
16. Collins, P. J., Ingram, G., and Codd, J. Simulated meltdown and vapour explosion experiments in ZEBRA 8G and ZEBRA 12 and their interpretation, Proceedings of the International Symposium on Physics in Fast Reactors, Tokyo, Oct. 1973.
17. Edwards, A. G., and Mather, D. J. The contribution of Bethe-Tait analysis to the assessment of fast reactor safety, Proceedings of International Conference on Engineering of Fast Reactors for Safe and Reliable Operation, Karlsruhe, October 1972, pp. 1287-1304.
18. Nicholson, R. B., and Jackson, J. F. A Sensitivity Study for Fast Reactor Disassembly Calculations, Argonne National Laboratory, ANL 7952 (1974).
19. Hunt, D. L., and Moore, J. G. Problems associated with molten fuel in LMFBR's (Chapter 4), in "An Appreciation of Fast Reactor Safety," UKAEA, 1970.
20. Hesson, J. C., Sevy, R. H., and Marciniak, T. J. Post-Accident Heat Removal in LMFBR's: In-Vessel Considerations, Argonne National Laboratory, ANL 7859 (1971).
21. Peckover, R. S. The use of core catchers in fast reactors, International Meeting on Reactor Heat Transfer, Karlsruhe, October, 1973, Paper No. 33.

22. Peckover, R. S., and Hutchinson, I. H. Thermal Convection Driven by Internal Heat Sources, UKAEA Report CLM-R123 (1973).
23. Jansen, G., and Stepnewski, D. D. Fast reactor fuel interactions with floor material after a hypothetical core meltdown, *Nucl. Technol. 17*, 85-95 (1973).

Index

A

Accidental releases of fission products, *see also* Fission products
  costs of, 58
  probability distribution of, 64
Accident conditions
  contaminant sprays and, 26–28
  environmental consequences of, 31–46
  filter systems and, 24–26
  fission product behavior in, 18–28
  passive systems and, 28
  reliability targets and, 111–112
  safety criterion for, 66
  for thermal reactors, 53
  thyroid cancer risk in, 54
Accidents, risk of death from, 50–52
Actinium series, 7
Activated carbon beds, methyl iodide and, 25
Afterheat, *versus* time, 110
Aircraft, risk from, 69
Alpha particle
  in radioactivity decay, 6
  stopping of, 8
Atomic energy, risk inherent in, 49, *see also* Nuclear power reactor
Automatic protective system
  failure characteristics of, 96–101
  for gas-cooled reactor, 84

B

Beta particle
  in radioactive decay, 6
  stopping of, 8
Boiling water reactor, *see also* Thermal reactor
  loss-of-pressure faults in, 179
  steam replacement in, 175
British Medical Research Council, 46

C

Calculated risk, 49–71, *see also* Risk
Can ballooning, in loss-of-pressure faults, 179
Cancer
  deaths for, 52
  radiation effects and, 34
  risks of, 54
Cesium-137, emergency reference levels of, 37
Chemical poisons, *versus* radioactive poisons, iii
CNEN Conference (1962), xii
Common fault modes, in heat removal reliability system, 135–138
Containment, hypothetical severe accident and, 201–206
Control rods
  failure of, 174
  hydraulic actuation of, 174

D

Dangerous equipment failures, defined, 86–87
Death
  occupation and, 52
  overall risk of, 50
  probability of from all types of accidents, 50
Decay heat, *see also* Heat removal systems
  versus decay time, 110–111
  in fast reactor, 122–123
  flow diagram for, 124
Decay heat rejection capacity for fast reactor, 125
Decay heat removal, 111
  failure probability for, 134–135
  for fast reactors, 120–122
  system failure criterion in, 130-131
Dripping water, fission products and, 23
Dry-out, in thermal reactor, 168

E

Electrical components, failure rates for, 76
Emergency cooling system, flow diagram of, 117
Emergency cooling water, loss-of-pressure faults and, 178
Emergency heat sinks, 115
Emergency reference levels
  in radiation dosage, 36–37
  for various isotopes, 37
Environmental consequences, of reactor accidents, 31–46
Equipment failure rates, 74–77, 89–91
  data collection in, 93–96
  observed *versus* predicted, 92, 94
Equipment failures, "dangerous" type, 86
Explosives, deaths from, 68
Extreme fault analysis, 87

F

Fail-dangerous conditions, 91
Fail-safe systems, 81
Fail-to-danger systems, 81
Failure probabilities
  of components and systems, 46–101
  versus reliability characteristics, 99
Failure rate
  characteristics of, 75
  determination of, 75
  for electronic components, 76

for protective equipment, 89–91
Fast reactor
  absorber removal in, 193
  component failure rate and repair data for, 156–157
  components of, 186
  containment in, 201
  coolant density changes in, 194
  coolant loss in subassembly of, 195–199
  core rearrangement in, 195
  data recording equipment for, 187
  decay heat rejection to condenser in, 122–123
  decay heat removal systems for, 120–122, 204–205
  delayed neutrons in, 188–189
  Doppler effect in, 191
  faults producing reactivity increases in, 193–195
  fire hazards in, 206
  fission product release in, 203–204
  fuel handling faults in, 199–200
  fuel vaporization in, 46
  heat removal in primary circuits of, 124–125
  hypothetical severe accident in, 201–206
  overall voidage effect in, 191–192
  plant flow diagram for, 119
  prototype, *see* Prototype fast reactor
  rapid insertion of fuel in, 193
  reactivity feedback processes in, 190–192
  safety of, 183–207
  sodium void effect in, 191
  sodium/water reactions in, 205–206
  sources of hazard in, 192–195
  special features of, 188–192
  specific power of, 189–190
  vapor explosion in, 197
Fault tree, for thermal reactors, 181
Filter systems
  contaminant sprays and, 26
  for gas cleaning, 24–26
Fires, deaths from, 68
Fission products
  "acceptable" frequency of release of, 60
  accidental releases of, 57–58
  behavior of in accident conditions, 18–28
  behavior of outside core, 17–18, 22
  behavior of in reactor operation, 13–18
  calculations for, 9–11
  chemical form of, 13–18
  condensation of, 22
  containment sprays and, 26–28
  costs of accidental release of, 58

Index 213

dripping water and, 23
escape of, 11
filter system for, 24–26
ground deposition of, 43–44
hazards for, 5–6
iodine-131 in, 39–41
isotopes in, 38–39
oxygen-to-metal ratio and, 15
path formations with, 14–15
quantities and relative importance of, 8–13
in reactor accidents, 37–46
release of from fuel, 14–19
strontium oxide in, 39
surface effects and, 12
thermal power and, 38
for thermal reactors, 53–54
thyroid cancer and, 56–57
type of fuel and method of manufacture in release of, 15–16
in uranium fuel, 12, 20–21
and vapor pressure on fuel can, 16
weather conditions and, 56
yields of for uranium and plutonium, 10
Fission yield, *versus* mass number, 9
Fuel pin, heat capacity of, 169

G

Gamma particles, stopping of, 8
Gamma radiation, from ground deposition, 45
Gamma rays
 in radioactivity decay, 6–7
 whole-body radiation for, 41
Gas-cooled reactor
 automatic protective system for, 84
 heat removal system for, 138
 loss-of-coolant faults in, 177–178
Genetic risk, 70–71
Geneva Conference on Peaceful Uses of Atomic Energy, xi, 40
Guard line, relay in, 83

H

Health physics control levels, radiation hazards and, 31–37
Heat rejection, basic methods of, 115–118
Heat removal system
 common fault modes in, 135–138
 component failure rate and repair data for, 128–129, 136–137
 criterion for system failure in, 130–131
 for gas-cooled reactors, 138
 immediate repairs to, 143–145
 logic diagram for, 126–130
 mathematical models of, 141–145
 NOTED program in analysis of, 131–133
 one- and two-parameter distributions for, 142
 for pressurized reactors, 138–140
 probabilistic theory application for, 140
 regular inspection and maintenance for, 142–143
 reliability of, 109–145
 results of analysis in, 133–138
 statistical failure distribution for, 141
 system failure probability to start in, 132
 for various reactor systems, 118–125
Heat sinks, emergency, 115
Heat sources, afterheat and, 110–111
Heavy water reactor, shutdown system for, 102
HTGCR reactor
 fission products in, 17
 methyl iodide and, 18
Human maloperation, in reactor system faults, 103

I

ICRP (International Commission of Radiological Protection), 31–33, 51
Instrumentation
 failure rates for, 76–77
 for protective systems, 78–86
 typical systems in, 77–78
"Involuntary" exposure, risk of, 62
Iodine-131
 deposition of in pastures, 43
 emergency reference levels for, 37
 in fission products, 39–40
 gas-borne concentration of fission products and, 23
 inhalation and absorption of, 42, 63, 67
 lethal and semilethal doses of, 63
 limits for release of, 150
 thyroid dose of, 42
Ionization, in radioactive decay, 8

K

Kaiser effect, in pressure vessel failure, 161
KITT computer program, 113

L

Leukemia, deaths from, 35, 50, 52
Liquefied natural gas explosion, 68
LMFBR
  damaged fuel in, 22–23
  fission products from, 17
LOFT test, 178
Logic diagram, for heat removal system reliability, 126–130
Logic sequence, in reactor protective system, 100
Loss-of-coolant accident, for pressurized reactors, 139
Loss-of-flow faults, 176–178
Loss-of-pressure faults, 178–182
Lung cancer, detection of, 35

M

Man-*versus*-machine comparisons, 104
Maximum credible accident, defined, xii
Maximum permissible radiation doses, 31–34
Meteorites, risk of bombardment from, 68–69
Methyl iodide, 18, 22–23
  activated carbon beds and, 25
Milk, iodine-131 in, 43

N

National Academy of Sciences, 35
Neutrino, in radioactive decay, 6
Neutron absorbers, in shutdown systems, 85
Niobium, vaporization and release of, 45
NOTED program, 113, 131–133
Nuclear industry, safety criteria for, 59
Nuclear installations, potential for harm from, xi
Nuclear power
  early development of, xi–xii
  risk inherent in, 49, 51
Nuclear reactor, *see also* Fast reactor, Reactor, Thermal reactor
  licensing of, 70
  protective requirement for, 78–81
  public anxiety about, 70–71
  quenching of by shutdown, 168
  risk from, *see* Risk
  risk/consequence relation in, 149
  safety circuits for, 82–84

O

Occupations, deaths as factor in, 52

P

Particulate filters, "absolute," 24
Passive systems, in accident conditions, 27–28
Plutonium-239
  alpha emission and, 45
  inhalation of, 45–46
Pony motors, in forced convection, 118
Potentiometer, open-circuit fault in, 89
Practical failure rates, 91–93
PREP computer program, 113
Pressure vessel
  integrity of, 149–165
  reliability requirements for, 152
Pressure vessel failures, 151–164
  categories of, 151
  causes of, 154
  cracks and, 155
  in design, materials, or construction, 159–160
  fail-danger rate of, 163
  German experience in, 156–157
  in Great Britain, 153–156
  Kaiser effect in, 161
  leakage in detection of, 162–163
  leak-before-break situation in, 163
  in nuclear applications, 158–159
  probability in, 164
  relevant experience in, 153–158
  statistical evidence in, 152–153
  testing in, 160–161
  ultrasonics in, 161–162
  US experience in, 157–158
  visual examination in, 162
Pressurized water reactor *see also* Thermal reactor
  heat removal system for, 138–140
  liquid boron poison in, 175
  loss-of-coolant accident in, 139
Probability of failure, *see* Failure probability
Protective channels, reactor fault condition in, 81
Protective equipment, reliability assessment of, 86–106
Protective requirement, for nuclear reactors, 78–81
Protective systems, *see also* Reactor protective systems

Index

analysis of, 78–86
failure characteristics of, 96–101
for gas-cooled reactors, 84
overall performance of, 86
protective channels and, 81–82
safety circuits and, 82–84
shutdown system in, 85
Prototype fast reactor, 119–123
schematic diagram of, 184

Q

Quantitative approach, to control reliability, 73–106

R

Radiation dose, *see also* Whole-body radiation dose
  cancer and, 34, 54
  contributors to, 51–52
  emergency reference levels in, 36–37
  maximum possible, 31–34
Radiation effects, 7–8, 31–46
  genetic, 35
Radiation hazards
  environmental consequences of, 31–46
  health physics control levels and, 31–37
Radiation protection, need for, 31
Radioactive cloud, iodine-131 inhalation from, 42, 63
Radioactive poisons, *versus* chemical poisons, iii
Radioactivity
  fission products and, 5–28
  effects of, 6–8, 31–46
Radioactivity decay, 6–7
Reactivity
  control rod removal in, 174
  defined, 172
  minimum required negative insertion of, 85
Reactivity faults, in thermal reactors, 172–176
Reactivity feedback processes, of fast reactors, 190–192
Reactivity increases, in fast reactors, 193–195
Reactor, *see* Fast reactor, Nuclear reactor, Thermal reactor
Reactor accidents, *see also* Accident conditions
  environmental consequences of, 31–46
  fission product release in, 37–46

Reactor control and instrumentation, schematic diagram of, 78, *see also* Instrumentation
Reactor fault condition, protective channels and, 81–82
Reactor hazards, evaluation of, xi, *see also* Risk
Reactor operation, fission product behavior in, 13–18
Reactor protective system
  assessment of overall performance in, 105–106
  block diagram of, 80
  failure characteristics of, 96–101
  human maloperation in, 103
  logic sequence diagram for, 100
  men *versus* machines in, 104
  operator and computer in, 103–105
  safety assessment in, 106
Reactor risks, compared with aircraft and meteorite risks, 69, *see also* Risk
Reactor safety, fields encompassed by, 2
Reactor safety circuits, 82–84
Reactor system components, failure probabilities of, 96–101
Reactor systems
  heat removal systems for, 118–125
  mean failure rate for, 96
Release frequency limit lines, 150
Reliability
  component failure rate and repair system in, 128–129
  computer programs for, 113–114
  of decay heat rejection, 123–125
  defined, 74, 112
  equipment failure rates and, 74–77
  of heat removal systems, 109–145
  quantitative approach to, 73–106
  probability distribution in, 74
  of pressure vessels, 152–160
  of protective equipment, 73–74
  safety circuits in, 82–84
  substantiating of, 114–115
  of thermal syphon loops, 125–126
Reliability analysis technique, 109
Reliability assessment, 112–118
  procedure in, 86–91
  system testing of predictions in, 114
Reliability target, accident conditions and, 111–112
Risk
  "acceptable" criterion for, 60–61
  acceptance of, xiii, 2

for aircraft, 69
calculated, 49
elimination of, 2
genetic, 70–71
"involuntary" exposure to, 62
of manmade accidents, 68
need for, 49
redistribution of, 49
reduction of, xiii
safety criterion and, 66
safety studies and, 66
Risk rate, calculation of, 54–55
Ruthenium-106
energy reference levels for, 37
inhalation of, 42
Ruthenium oxide, inhalation of, 42–43

S

Safeguards, "high dependability" in, xii
Safety assessment, steps in, 106, *see also* Reactor protective system
Safety circuits, 82–84
Safety criteria, development of, xii
SAMPLE computer program, 113
Shutdown
heat sinks in, 116
quenching by, 168
Shutdown systems, 85–86, *see also* Reactor protective system
analysis of, 101–103
Sodium hydroxide, in contaminant sprays, 27
Sodium iodide, in LMFBR reactor, 18
Sodium thiosulphate, in containment sprays, 27
Sodium void effect, 191
Sodium/water reactions, in fast reactor, 205–206
Spray systems, 26–27
Steam generator, design of, 185
Strontium, volatility of, 11
Strontium oxide, in fission products, 39
Surface effects, fission products and, 12

T

TEDA, *see* Triethylenediamine
Temperature trip amplifier, 88
failure rates for, 90
Thermal reactor
accidents to, 53–54, 176
control rod ejection accident in, 176
control rod withdrawal rates for, 170
dry-out in, 168
fault tree for, 181
fission product releases in, 171–172
fuel pin temperature distribution for, 168–169
local overheating in, 170
loss-of-flow faults in, 176–178
loss-of-pressure faults in, 178–182
meltout of fuel in, 172
reactivity faults in, 172–176
scram conditions in, 168–171
thermal inertia of fuel in, 167
Thermal reactor safety, 167–182
Thermal syphon loops, reliability of, 125–126
Thorium series, 7
Thyroid cancer, 35
deaths from, 50
fission product release and, 56–57
risk rates for, 54, 67
Thyroid gland, iodine inhalation and, 42, 63
Transient bottom duct fracture, 79
Triethylenediamine, 26
Trip amplifier, 88

U

United Kingdom Atomic Energy Authority, 3, 45, 52
United Kingdom Medical Research Council, 36
Uranium dioxide fuel, measured values of fission product release from, 20–21
Uranium fuel, activities of fission products in, 12
Uranium series, 7
US Federal Radiation Council, 36–37

V

Vapor explosion, in fast reactor, 197
Vaporized fuel, release of, 45–46

W

Washout coefficient, 27
Weather condition, probability distribution of, 64
Whole-body radiation, cancer from, 54
Whole-body radiation dose, 31–35, 41, 51
"permissible" level of, 71

Z

Zirconium, vaporization and release of, 45